TOPICS IN

ASTROPHYSICS AND SPACE PHYSICS

EDITED BY

A.G.W. CAMERON and GEORGE B. FIELD

V. L. GINZBURG--Elementary Processes of Importance for
Cosmic Ray Astrophysics

V. L. GINZBURG--The Origins of Cosmic Rays

THE ORIGIN OF COSMIC RAYS

Vitalii Lazarevich

V. L. Ginzburg

with

S. I. Syrovatskii

P. N. Lebedev Physical Institute
Academy of Sciences of the USSR
Moscow

GORDON AND BREACH, SCIENCE PUBLISHERS

New York / London / Paris

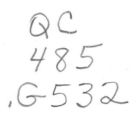

Editorial office for Great Britain:
Gordon and Breach, Science Publishers Ltd.
12 Bloomsbury Way
London W.C.1

Editorial office for France:
Gordon & Breach
7-9 rue Emile Dubois
Paris 14e

Distributed in Canada by:
The Ryerson Press
299 Queen Street West
Toronto 2B, Ontario

Part I is reprinted, with additions and corrections, from DeWitt, Schatz-
man, Véron, Eds., *High Energy Astrophysics:* Lectures delivered at Les
Houches during the 1966 session of the Summer School of Theoretical
Physics, Vol II, Gordon and Breach, 1967. Part II was translated from
Proc. Tenth Intern. Conference on Cosmic Rays, Part A, p. 48, Calgary,
Canada, 1967.

Printed in the United States of America   (0197)

PREFACE

This book is based on a lecture course prepared by myself in 1966 for the summer school at Les Houches. The course was entitled: "The Origin of Cosmic Rays."

Since the author was unable to deliver the courses, the text of the lectures is merely the English translation of the original Russian lectures which were sent to Les Houches. The translation has been examined by the author, and a number of minor clarifications have been made. There was no attempt at basic change or supplementation. Therefore, with the aim of reflecting new material, one article is being added:

V. L. Ginzburg and S. I. Syrovatskii, "On the Origin of Cosmic Rays," *Proc. Tenth Intern. Conference on Cosmic Rays,* Part A, p. 48, Calgary, Canada, 1967.

I would like to express my gratitude to the publishers and more concretely to R. Akerib for the great attention shown in preparing this book for publication.

V. L. Ginzburg

CONTENTS

PART I

INTRODUCTION

What are the main constituents of the Universe? Some time ago one would have considered only stars, gas, optical radiation, solid components (dust, planets). Nowadays, one must undoubtedly add cosmic rays (various high-energy particles). On the one hand, this conclusion comes from the discovery of cosmic rays in a great number of objects (supernovae envelopes, galaxies, radiogalaxies, and quasars), and on the other hand from the clarification of the role of cosmic rays as a dynamical and energetic factor.

Understanding the important role of cosmic rays in the Universe is one of the major events of modern astrophysics. Thus, taking into account the special nature of the topic, it is convenient to distinguish a branch of astrophysics—the astrophysics of cosmic rays. This topic, though usually called "the origin of cosmic rays" is more general than that title implies.

Information on cosmic rays in the Universe arises from a whole series of investigations:

From the study of the chemical composition, energy spectrum, and other characteristics of primary cosmic rays incident on the Earth.

From radioastronomy methods—taking into account the fact that the nonthermal radiation is synchrotron (magnetobremsstrahlung) radiation from relativistic electrons and positrons.

From optical methods—optical synchrotron radiation, and indirectly by using various results from optical astronomy.

From γ and x-ray astronomy (until now little has been obtained with these methods but the prospects are very great).

The study of meteorites is of well-known importance, it gives information on the intensity of cosmic rays in the past (as far back as 10^9 years ago) but strictly speaking only in the solar system.

Finally, data from neutrino astronomy is a future possibility.

In conclusion, the development of cosmic-ray astrophysics is going by different ways, using results from many sources, and furthermore the possibilities of performing observations is rapidly increasing. The following phenomena are studied: solar events, which are related to the cosmic rays in the solar system and on the Earth (radiation belts*), galactic cosmic rays at the Earth, cosmic rays in the Galaxy and in the envelopes of galactic supernovae, in other normal galaxies, radiogalaxies, and quasars.

---

*The question of the radiation belts has more connection with geophysics than with astrophysics; but here such a problem of terminology is unimportant.

It is clear that radioastronomy and optical data, while related to cosmic rays at distances far from the Earth give direct information only on the electronic component. One must also emphasize the fact that the problem of intergalactic cosmic rays is of great interest: but there are almost no direct data as yet (see later).

Information about cosmic rays can be and actually is used with other data for the analysis and understanding of a whole series of problems: supernovae and radiogalactic explosions, structure of quasars, formation and structure of the galactic halo, etc . . . .

On the other hand there are problems related to the cosmic rays near the Earth, acceleration mechanisms, etc . . . .

One can see that cosmic-ray astrophysics is by now a well developed science and that on account of the whole material covered would require more than one book. The book[1] on which we shall rely has 400 pages and 500 references, but covers at most only half of the subject. Reference 2 is devoted to solar cosmic rays and related problems. For details we shall refer to these two books and to the Proceedings of the last two conferences on cosmic rays (Ref. 3 and 30). The aim of the present lectures is to convey some feeling for cosmic-ray astrophysics without going into too much detail*. At the same time we shall try to emphasize the major problems, those that are not yet resolved, so that the reader may judge the state of present-day cosmic-ray astrophysics.

I. COSMIC RAYS NEAR THE EARTH

I. 1. PRELIMINARY REMARKS

Studying primary cosmic rays near the Earth is difficult, for one has to send the apparatus (photo emulsion, various counters, cameras) to high altitudes. There must be no more than a few g/cm^2 of matter (air plus any part of the equipment) above the apparatus. (The whole atmosphere is 1000 g/cm^2 thick.) This requirement can be met by high-altitude balloons (altitude ~30–40 km), but in this case it is hard to ensure long exposure times (this remark applies to rockets also). In the case of satellites difficulties arise in getting sufficiently detailed information and furthermore, in a number of experiments, in knowing the orientation of the apparatus. Note that many events one would like to register, such as the passage of heavy or very high energy particles, are very rare. Even the measurements for which the influence of the overlaying air layers can be neglected, are somewhat complicated to interpret because of the influence of the Earth's magnetic field (not to mention such perturbations as the "albedo", and the particles of the radiation belt). Finally, when galactic cosmic rays are measured near the Earth (outside the magnetosphere), some corrections must be made for interplanetary matter to distinguish them from solar cosmic rays. These remarks show why, in spite of the efforts of the last few years, we do not have very complete data on the primary cosmic rays near the Earth†.

As a result, it would be difficult to present a complete series of results together with their degrees of accuracy.

We shall make no such attempt and shall only give indicative results (for further information, see Ref. 3, 5, 28 and 30.

It is clear that there is not, generally speaking, sufficient data to build a

---

*Note that a whole series of data on elementary processes important in cosmic-ray astrophysics is given in Ref. 4 (see this book, page 325). Data included in Ref. 30 were not included in this chapter written in 1966. (See, however, the next paper, page 41)

†We would like to emphasize that from now on we have all the means to make valid progress in the study of primary cosmic rays, mainly with satellites; so we may hope that the coming years will bring the answers to a number of questions which are still unclear.

quantitative theory of the origin of cosmic rays. Furthermore the results of quantitative computations depend not only upon information on primary cosmic rays at the Earth, but also on many other data that are not so very well known.

We can make this clear with the example of the chemical composition of cosmic rays at their source. In order to solve this problem, we must know the following things:

> the chemical composition of the cosmic rays at the Earth, the probabilities of various nuclear reactions (the reactions taking place in the interstellar medium), the distances of the sources, the particle trajectories in the Galaxy, etc. . . . .

Therefore it is easy to see that data on the chemical composition of primary cosmic rays do not suffice to determine the composition at the source, or the average thickness of the matter crossed by the cosmic rays in the interstellar medium. Obviously one can only sketch the features of a quantitative theory (see Ref. 1, Chap. 5), and one has to start mostly with qualitative or semiquantitative computations. This is the method we shall use, and therefore the degree of accuracy of the data in most cases is of little importance.

What we have said should not be understood as a pessimistic opinion of either the importance or the possibility of a quantitative theory for the origin of cosmic rays, nor as a rejection of the possibility of obtaining and analyzing more precise data. Obviously, building up a quantitative theory must be the next step in the development of the Astrophysics of cosmic rays. Strictly speaking, we have already started this work and it has proved hopeful and interesting. But what is now the essential point, as we have said, is the qualitative picture and choice of models.

It is clear that here we must focus our interest on these problems.

The fundamental quantity which characterizes the cosmic rays is their intensity I (sometimes this quantity is called the flux in a given direction). By definition I is the number of particles incident per unit time on a unit area at right angles to the direction of observation, with respect to a unit solid angle. It is measured in:

$$\frac{\text{Number of particles}}{\text{cm}^2 \text{ ster sec}} = 10^4 \frac{\text{Number of particles}}{\text{m}^2 \text{ ster sec}}$$

The flux of particles of type i for which the intensity is equal to I_i is:

$$F_{i, \Omega} = \int_\Omega I_i \cos \theta \, d\Omega,$$

where θ is the angle between the vector perpendicular to unit area and the velocity of the particles, and $d\Omega$ is an element of solid angle. For isotropic radiation, the flux of particles F_i directed into a whole hemisphere is equal to

$$F_i = 2\pi \int_0^{\pi/2} I_i \cos \theta \sin \theta \, d\theta = \pi I_i \tag{1.1}$$

In the case of isotropic radiation, the density N_i of particles with velocity v_i is equal to

$$N_i = \frac{4\pi}{v_i} I_i \tag{1.2}$$

Generally we do not deal with monoenergetic particles but with particles which have a certain energy distribution (i.e. energy spectrum), in this case the essential quantity in the spectral (differential) intensity $I_i(E)$, defined such that $I_i(E)dE$ is the intensity of particles with total energy between E and E + dE.

The intensity of particles with energy $> E$, (integral intensity) is equal to

$$I_i(> E) = \int_E^\infty I_i(E)\,dE \tag{1.3}$$

For an isotopic distribution of particles with mass M_i:

$$N_i(> E) = 4\pi \int \frac{I_i(E)}{v}\,dE, \quad E = \frac{M_i c^2}{\sqrt{1 - v^2/c^2}} \tag{1.4}$$

In practice, we shall only have to deal with isotropic particles because the degree of anisotropy of cosmic rays is very low (less than 1%, cf. I.3). Thus, unless explicitly stated otherwise, we shall always deal with an isotropic distribution of particles. The aim of the measurements is of course the determination of the spectrum $I_i(E)$ for all components (index i) of the cosmic radiation. In practice, the problem is split up into two parts:

To find the chemical composition $I_i(E)/I_j(E)$ or $I_i(>E)/I_j(> E)$ for various nuclei or groups of nuclei (indices i, j), and to find the energy spectrum $I_i(E)$ for protons, for nuclei, for electrons or for all cosmic rays together. A particular problem is the determination of the anisotrpy, i.e. the dependence of $I_i(E, \theta)$ on θ in any arbitrary direction.

I.2 PROTONS AND NUCLEI (CHEMICAL COMPOSITION)

The fundamental components of cosmic rays near the Earth are protons and nuclei (the intensity of the electrons and positrons is only ~1%, of the total, cf. Sec. I.4).

To exhibit the characteristic features of the chemical composition of the galactic cosmic rays near the Earth, it is convenient to use the integrated intensity, choosing a sufficiently high threshold energy. (In the low-energy range one has solar effects which depend on the solar activity.*)

TABLE I. Chemical composition of cosmic rays.

Nucleon group	Z	$\overline{A}$	Intensity $I\ (>\epsilon = 2.5$ BeV/nucleon) $(m^{-2} sr^{-1} sec^{-1})$	Number of nucleons in the flux	$\dfrac{I}{I_H} = \dfrac{N}{N_H}$	Average in the universe according to different data	
p	1	1	1300 ± 100	1300	650	3360	6830
α	2	4	94 ± 4	376	47	258	1040
L	3-5	10	2.0 ± 0.3	20	1.0	10^{-5}	10^{-5}
M	6-9	14	6.7 ± 0.3	94	3.3	2.64	10.1
H	≥ 10	31	2.0 ± 0.3	62	1.0	1	1
VH	≥ 20	51	0.5 ± 0.2	25	0.26	0.06	0.05
VVH	≥ 30		$\sim 10^{-4}$	Total 1880	$\sim 10^{-4}$	$\sim 10^{-4}$	

*The low-energy range ($\epsilon < 1$-2 BeV/nucleon) is very interesting, and is now studied intensively, but we shall leave it aside in the present lectures, together with other problems connected with solar activity (see Refs. 2, 3, and 5).

In Table I we give the chemical composition of cosmic rays with total energy $\epsilon = E/A \gtrsim 2.5$ BeV/nucleon, where E is the total energy of the particle of mass number A. To this energy, for particles of atomic number $Z \gtrsim 2$, there corresponds the vertical rigidity at the threshold $R_H = pc/eZ = p(BeV/c)/Z = 4.5$ BeV (p is the momentum of the particle, eZ is the charge). Such a rigidity corresponds to a geomagnetc latitude $\lambda_m = 41°$ (Texas, Northern Italy). For protons, $Z = A = 1$, and for a total energy $E = \epsilon = 2.5$ BeV, the rigidity is equal to $R_H \simeq 2.3$ BeV. Even for larger rigidities ($R_H = 4.5$ BeV) the intensity of the cosmic rays is sensitive to the degree of solar activity. (During maximum, the intensity is smaller than during minimum by 20 - 30%.)*The data given in Table I correspond to the period of minimum activity, during which the data are most similar (but not exactly identical) to those of interstellar space near the solar system.*

There are some comments on Table I in Ref. 1 and Ref. 28. Here we shall be satisfied with a few remarks. Following normal practice, data are shown in Table I for protons and groups of nuclei. (The α particles consist of He^3 and He^4; group L consists of Li, Be, B; group M of C, N, O, F; group H of nuclei with $Z \geq 10$; sub group VH of nuclei with $Z \geq 20$). The presence of deuterium in relativistic primary cosmic rays has not yet been established. One can conclude that the deuteron intensity is not larger than 2-5% of the proton intensity. The presence of He^3 in cosmic rays can be considered as certain and the flux ratio is

$$\frac{He^3}{He^3 + He^4} = 0.1\text{-}0.2.$$

This is taken from data for helium nuclei with energies between 80-350 MeV/nucleon (see Ref. 3). For relativistic helium nuclei, the ratio $He^3/(He^3 + He^4)$ is not clearly established in spite of its great interest (see Ref. 7). Data now exists, not only on the groups of nuclei, but also on individual nuclei within the groups. But we cannot make full use of these data because of a lack of sufficient knowledge of the spallation probabilities. Therefore we shall restrict ourselves to two basic conclusions obtained from Table I.

There are about five orders of magnitude more nuclei (Li, Be, B) per H nucleon in cosmic rays than for the cosmic average. (The data are in the last two columns of Table I.) It seems very unlikely that the chemical composition of the cosmic-ray sources should be so anomalous. It is at the same time absolutely unavoidable that some L nuclei are produced as a result of the spallation of heavier nuclei in the interstellar medium. Then is it reasonable to suppose that in practice all Li, Be, and B nuclei, and other rare nuclei (such as He^3) appear in the cosmic rays as a result of the spallation of heavier nuclei. Therefore, it appears possible to obtain information about the path of the cosmic rays through the interstellar medium, and about their composition at the source. The corresponding computation is given in Ref. 1, Sec. 15. Some recent results and references are given in Refs. 6 and 7. The most simple and the most important parameter upon which the ratio $\frac{L}{(M + H)}$ depends is the average thickness x of the matter crossed by the cosmic rays between their source and the Earth. The value of x depends rather weakly upon the models and exact parameters of the spallation. To the observed ratio $L/(M + H) \simeq 0.25$, there corresponds a thickness $x = 2\text{-}10$ g/cm². We shall assume that x is equal to 3 g/cm². If the average density of the matter on the cosmic-ray path is $\rho \sim 2 \times 10^{-26}$ g/cm³ (i.e. the particle density of the interstellar gas is $n \sim 10^{-2}/cm^3$), we obtain the mean path length $L = x/\rho \simeq 2 \times 10^{26}$ cm, and the time of travel

---

*The degree of precision is not yet clear; one can think that for $R_H \geq 4.5$ BeV the difference does not exceed 5-10%, but for $R_H \geq 2.5$ BeV it can reach 30-50%.

$$T = L/c \sim 6 \times 10^{15} \text{ sec} = 2 \times 10^8 \text{ years.} \qquad (1.5)$$

The distance from the center of the Galaxy to the Sun is about 10 kpc = 3×10^{22} cm, and the characteristic radius of the Galaxy is R $\simeq$ 15 kpc $\simeq 5 \times 10^{22}$ cm. Thus, the ratio R/L $\sim 10^{-4}$. In principle, no difficulty arises here for the galactic model of the origin of the cosmic rays.* As a matter of fact, the cosmic rays are strongly deflected by the magnetic fields H $\sim 10^{-6}$-10^{-5} oersted, which exist in the Galaxy, according to a number of data. The radius of gyration of an ultrarelativistic particle with energy E and charge eZ moving in a plane perpendicular to H is equal to

$$r = \frac{c}{\omega_H} = \frac{E}{eZH} = \frac{E(eV)}{300ZH}, \qquad \omega_H = \frac{eHc}{E}. \qquad (1.6)$$

For Z = 1 and H = 10^{-6} oersted,

$$r(\text{cm}) = 3 \times 10^3 \text{ E(eV)}.$$

It is obvious that one has

$$r \sim 10^{-2} R \sim 3 \times 10^{20} \text{ cm}$$

even for E $\lesssim 10^{17}$ eV.

Since the cosmic-ray energy is E $\lesssim 10^{10}$ eV for the majority of cosmic rays, one has r $\ll$ R. Apart from this, the characteristic length l for the nonuniformity of the magnetic field is probably also small compared to R. In fact, there exists in the Galaxy a regular large-scale field H, but probably it is closed or nearly closed. In any case the assumption that the cosmic rays are accelerated and travel in the galactic field does not contradict the known data. In such a galactic model, we can conclude from the data on chemical composition that the time of travel of the cosmic rays is T $\simeq 2 \times 10^8$ years. [See (1.5) and Refs. 1 and 3.†]

The second conclusion to be taken from Table I is that the cosmic rays are much richer in heavy elements (relative to H and He) than the cosmic average. The relative abundance of very heavy nuclei (Z $\gtrsim$ 20) is 30 times the cosmic average. In the sources of the cosmic rays, there must be even more VH nuclei since some of them are destroyed before they can reach the solar system. Besides, some of the α particles and protons observed at the Earth are known to be secondary. (They are formed in the interstellar medium as a result of spallation.) Unfortunately the evaluation of the chemical composition and of the probabilities of spallation is not yet sufficiently accurate to allow a precise determination of the chemical com-

---

*We are considering the cosmic rays observed on the Earth. By the galactic theory, we mean the theory in which the cosmic rays are accelerated mainly in the Galaxy, (See Sec. III.)

†As we have seen, the evaluation of the time T is carried out using the data for the thickness x of the gas traversed and an assumption about the average density ρ of the gas in the region crossed by the cosmic rays (T $\sim$ x/ρc). It is also possible to compute the time T directly by measuring the abundance of the radioactive isotope Be^{10}. The lifetime of this isotope is $\tau_{1/2,Be} = 4 \times 10^6$ E/Mc^2 years. If the cosmic rays travel during a time T $\gtrsim \tau_{Be}$ then a fraction of the Be^{10} decays, but if T $\ll \tau_{Be}$, then the number of Be^{10} nuclei is proportional to the probability of their formation. According to preliminary data (Ref. 8), one finds T $> 5 \times 10^7$ years. Knowing x and T, from independent sources one can evaluate ρ = x/cT; for x = 3 g/cm^2, T $\gtrsim 5 \times 10^7$ years, $\rho \simeq 2 \times 10^{-24}$ n, one has $\rho \lesssim 6 \times 10^{-26}$ g/cm^3 and n $\lesssim$ 0.03/cm^3 (New data see in Ref. 30; there are yet no realistic estimate of T from Be^{10}).

position of the cosmic rays at the source. In any case the difference observed between the chemical composition of the Universe and that of the cosmic rays (as much as two orders of magnitude in the extreme cases) is an important result. There are two possible explanations. The first is the assumption of a preferential acceleration for heavy nuclei—such a preferential acceleration is known to exist (Ref. 1, Sec. 9). The other assumption is that the sources of the cosmic rays are anomalous, i.e. very poor in hydrogen and helium; such an assumption cannot be excluded, even though it seems somewhat artificial.

Let us emphasize that the available data on the chemical composition correspond to the particles with energies less than 10^{12}-10^{13} eV. These particles whose energies are not so very high form the bulk of the cosmic rays. Nevertheless one observes particles of energies up to 10^{20} eV, and the problem of the chemical composition of the cosmic rays with energies $E > 10^{12}$-10^{13} eV and mainly $E > 10^{15}$ eV is interesting and important. However, there are no reliable data in this range. One may think that up to energies of 10^{15} eV the chemical composition is almost the same as for lower energies (see Table I); for $E > 10^{15}$ eV large changes in the chemical composition are possible. Unfortunately, progress has been insignificant in this range.* On the contrary, a whole series of results (see Ref. 3) about the chemical composition in the low-energy range ($E \le 1$ BeV/nucleon) has been gathered recently. Even here the picture is not yet clear, particularly in the field in which we are interested, i.e. the study of the galactic cosmic rays outside the solar system. Let us note that the ratio L/M of light to medium nuclei increases when the energy decreases down to $\epsilon = 400$ MeV/nucleon, but then it starts to decrease (see Ref. 3, p. 407-462). In the range of energies $\epsilon < 400$ MeV/nucleon the solar modulation is important. In this range, the ionization losses increase; at the same time the increase of L/M from 0.3 for $\epsilon \ge 2$-3 BeV/nucleon up to 0.5 for $\epsilon \simeq 400$ MeV/nucleon can probably be considered as an argument against a strongly nonstationary model for the origin of the cosmic rays in the Galaxy (see Ref. 6 and Sec. III.2 below.) Though this argument is short, we hope we have made it clear that the study of the chemical composition of the cosmic rays is full of possibilities. The means of determining the chemical — or even isotopic composition—of the primary cosmic rays are promising. The probabilities of spallation in cosmic rays can be determined more accurately using accelerator experiments. It is probable that soon the only unknown factors remaining will be the properties of the interstellar medium, the motion of the cosmic rays across it, and the nature of the sources. Under these conditions we will be able to choose, according to the data on the chemical composition, the models for the propagation of the cosmic rays and the initial composition (see Refs. 1 and 7). Along these lines we can expect to get results which will be important not only for cosmic-ray astrophysics but also for other parts of astrophysics.

I. 3. PROTONS AND NUCLEI (ENERGY SPECTRUM AND ANISOTROPY). INTEGRATED VALUE FOR THE PARTICLE INTENSITY AND ENERGY DENSITY OF THE COSMIC RAYS

The dependence of the intensity upon the energy (energy spectrum) is usually expressed in the following way:

---

*High-energy particles with $E > 10^{14}$ eV can be studied at the moment only by the atmospheric showers. Thus information on primary particles is indirect. For instance, if the shower is produced by a photon or an electron, it will be poor in μ mesons. The showers produced by protons and nuclei are different in many respects—spatial distribution and fluctuation of their various components. With these properties, one may hope to find an approximate solution to the problem of the chemical composition of the very high energy cosmic rays.

$$I_A(> \epsilon) = \int_\epsilon^\infty I_A(\epsilon)\, d\epsilon = K_A \epsilon^{-(\gamma-1)},$$

$$I_A(\epsilon) = (\gamma - 1)K_A \epsilon^{-\gamma}, \tag{1.7}$$

where $\epsilon = E/A \equiv$ energy per nucleon and the subscript A shows that we are considering nuclei (or groups of nuclei) with average mass number A. In addition, we introduce analogous quantities $I(> E)$ and $I(E)$ for all the cosmic rays taken together (E—particle energy).

As a matter of fact the spectrum is not a power-law spectrum since γ is a function of the energy. However, and this fact is important, in quite a large range of energies the approximation (1.6) is sufficiently accurate. Thus, in the range $2 \times 10^9 < E < 10^{15}$ eV/nucleon $\gamma = 2.5 \pm 0.2$ (or $\gamma = 2.7 \pm 0.2$ for some data). We shall take $\gamma = 2.6$. In fact, in this approximation, γ is the same for protons and other nuclei. For example, the spectrum is of this form for all cosmic rays taken together in the range $10^{10} < E < 10^{15}$ eV:

$$I(> E) = (5.3 \pm 1.1) \times 10^{-6}\left(\frac{E}{6 \times 10^{14}}\right)^{-(\gamma-1)} \text{ particles/m}^2\text{-sr-sec},$$

$$\gamma = 2.62 \pm 0.05, \tag{1.7a}$$

where E is in eV. In the low-energy range $E < 10^9 - 10^{10}$ eV the exponent γ changes and the spectrum depends strongly on the degree of solar activity. As already mentioned, we shall not consider this range (see Refs. 2, 3, and 5). Let us note that there is a very important problem (as yet unsolved) for theories of cosmic rays—namely that of the shape of the spectrum at low energies and that of its maximum (outside the solar system). Clearly, up to energies of $\epsilon \sim 100$ MeV/nucleon there is no maximum in the spectrum of the galactic cosmic rays.

At the energy $E \sim 10^{15}$ eV there is a more or less sharp change in the slope, or at least some change in the spectrum slope, and for $E > 10^{15}$ eV, the expression (1.7) is not valid. The spectrum which better describes reality for $E > 10^{15}$ eV is the following:

$$I(> E) = (2.0 \pm 0.8)\, 10^{-6}\left(\frac{E}{10^{15}}\right)^{-\gamma-1} \text{ particles/m}^2\text{-sr-sec},$$

$$\gamma = 3.2 \pm 0.2, \tag{1.8}$$

where E is in eV. For $E \sim 10^{15}$ eV the spectra (1.7a) and (1.8) agree within experimental errors. According to the best data known, for 10^{15} eV $< E < 10^{18}$ eV, one has the exponent $\gamma = 3.1 \pm 0.1$. It is possible that for $E > 10^{18}$ eV, the spectrum becomes flatter ($\gamma \simeq 2.6$). However this hypothesis has not yet been verified. The spectrum of the cosmic rays for $E > 10^{10}$ eV is shown in Fig. 1.

Notice that in the range where the chemical composition of the cosmic rays is unchanged, the particles of energy E (or of energy $>E$) are mostly nuclei.*

---

*This is connected with the decreasing character of the spectrum and its dependance on ϵ alone [see (1.6)]. Indeed, according to (1.6)

$$I_A(> E) = K_A\left(\frac{E}{A}\right)^{-\gamma+1} = K_A \frac{A^{\gamma-1}}{E^{\gamma-1}}.$$

It follows that

$$\frac{I_{A_2}(> E)}{I_{A_1}(> E)} = \frac{K_{A_2}}{K_{A_1}} \times \frac{A_2^{\gamma-1}}{A_1^{\gamma-1}},$$

while

$$\frac{I_{A_2}(> \epsilon)}{I_{A_1}(> \epsilon)} = \frac{K_{A_2}}{K_{A_1}}.$$

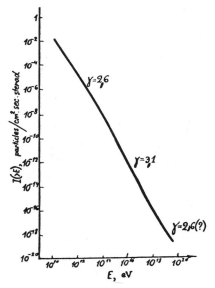

FIG. 1. Integral cosmic-ray energy
spectrum [in the $E \geq 10^{18}$ eV energy
range γ cannot be considered to be finally
established for the power-law integral
spectrum I ($>$ E) = const $E^{-(\gamma-1)}$].

So the ratio of the intensity of all nuclei with $Z \geq 2$ to the intensity of all cosmic rays at a given energy per nucleon is 7% (See Table I, where one can find the
values of $I_A (> \epsilon = 2.5$ BeV/nucleon)). Then, for protons one has $I_p/I_{total} = 0.93$.

On the contrary, the ratio of the number of protons with energy $>$E to the
total intensity of the cosmic rays with energy $>$E is 0.49 for $\gamma = 2.5$ and only
0.37 for $\gamma = 2.7$. For further details, see Table 5 in Ref. 1. Unfortunately, the
chemical composition of the cosmic rays in the range of energies higher than
10^{12}-10^{15} eV is not yet well known—particularly for $E > 10^{15}$ eV. It is quite possible
that in this range of energies (10^{15}-10^{17} eV) the metagalactic cosmic rays become
important (considering the cosmic rays observed near the Earth). The knowledge
of the composition of these cosmic rays is interesting from several points of view.
On the other hand, if all cosmic rays (for $E > 10^{15}$-10^{17} eV) were composed of VH
nuclei (which is very unlikely), then they might be of galactic origin.

Another interesting question is the character of the cosmic-ray spectrum for
$E > 10^{20}$ MeV. Even the strongest radiogalaxies could scarcely produce particles
with energies $>10^{22}$-10^{23} eV. (One can consider the difficulties of keeping the
particles in the accelerating regions, etc.). Therefore we could expect for some
energy $>10^{20}$-10^{21} eV a sharper decrease of the spectrum (increase of γ) such that
there is practically a complete cutoff.*

However, even at lower energies, a cutoff is produced by the creation of π mesons

---

*Note that even when there is no cutoff with extrapolating data, in the range
$E > 10^{20}$ eV, the number of particles in this region is very small. According to
Ref. 9 one has $I(E > 2 \times 10^{20}$ eV) $\sim 3 \times 10^{-3}$ particles/km^2-year. Nearly the same
result for I can be computed from formula (1.8).

in the collisions of protons with thermal photons that fill the intergalactic space. At our epoch, the temperature of these photons is about 3 °K,[*] corresponding to an energy density $w_T = 0.4$ eV/cm^3 and a mean photon energy of $\epsilon_T = 7 \times 10^{-4}$ eV. Photons with an energy $\epsilon_T = 7 \times 10^{-4}$ eV produce π mesons during collisions with protons of energies

$$E > E_c \simeq \frac{m_\pi c^2}{2\epsilon_T} \times M_p c^2 \sim 10^{20} \text{ eV}.$$

Due to this effect (for $E > 10^{20}$ eV) the spectrum of the protons can be considered as practically cutoff (see Ref. 9). In the case of nuclei, an essential phenomenon is the photodisintegration which leads also to a cutoff in the spectrum for energies $E > 1 - 2 \times 10^{20}$ eV.

The optical thermal radiation ($\epsilon_{opt} \sim 1$ eV) in the Metagalaxy causes disintegration of heavy nuclei with energies $E > 10^{16}$ eV, but the energy density of this radiation is smaller. As we have already mentioned, $w_{opt} \sim 10^{-2}$ eV/cm^3; hence, only a part of the nuclei will decay.[†]

Apart from the chemical composition and energy spectrum of the cosmic rays, in principle another important feature is the degree of anisotropy.

$$\delta = \frac{I_{max} - I_{min}}{I_{max} + I_{min}} , \qquad\qquad (1.9)$$

where I_{max} and I_{min} are the maximum and minimum values of the intensity found by varying the direction. If the intensity of the cosmic rays in the analysed range depends only upon an angle θ relative to some Z direction, then in the simplest case:[‡]

$$I(\theta) = \bar{I}(1 + \delta \cos \theta), \quad I(0) = I_{max} = \bar{I}(1 + \delta),$$

$$I(\pi) = I_{min} = \bar{I}(1 - \delta) \text{ and } \bar{I} = \frac{1}{2}(I_{min} + I_{max}).$$

One important fact is that the cosmic rays are isotropic to a high degree, i.e. $\delta \ll 1$. Moreover the presence of an anisotropy cannot be considered at the present time to have been established: one has $\delta < 1\%$ (for $E > 10^{16}$ eV, the upper limit of δ is slightly larger).

Only during the years of minimum solar activity (1954), have we obtained some information (see Ref. 2) on an anisotropy up to a few tenth of 1%. The certainty of even a small degree of anisotropy would be important; it would be important, in particular, to know the direction in which the intensity has a maximum. The escape of the cosmic rays from the Galaxy should lead to some anisotropy (with $\delta \sim 10^{-3}$, see Ref. 1, Sec. 16), as the flux is directed from the center of the Galaxy to its outer

---

[*]We neglect here the metagalactic optical radiation for which the energy density is $w_{opt} \simeq 10^{-2}$ eV/cm^3 (see Refs. 1 and 4).

[†]The average lifetime of a nucleus against photodisintegration is

$$\tau = 1/\sigma n_{opt} c,$$

where the cross section for the photodisintegration is $\sigma \lesssim 10^{-25}$ cm^2, and the particle density of the optical photons is

$$n_{opt} = w_{opt}/\epsilon_{opt} \sim 10^{-2} \text{ cm}^{-3}.$$

From this, one gets $\tau \lesssim 10^9$ years and so the nuclei produced during the last 10^9 years have not been destroyed.

[‡]Evidently, when $\delta \ll 1$ we use an expansion in spherical harmonics with only the first term of the series.

regions. The magnetic fields near the solar system can naturally change the direction of the flux, but it is unlikely that the angle of rotation is greater than $\pi/2$ (i.e. that the flux would change its sign and correspond to the sign for the flux of metagalactic particles entering the Galaxy). Then the result that cosmic rays flux are directed towards the outer regions of the Galaxy (data given in Ref. 2 indicate that such is the case*), is an argument for the galactic origin of the cosmic rays (see Sec. III later). The high degree of isotropy of the cosmic rays probably is a plasma effect and has numerous implications (see Refs. 6 and 10 and Sec. III.4).

In the computation of the particle density of the cosmic rays, and of the relevant energy density, the cosmic-ray distribution can be taken as isotropic. [Use formula (1.2).] During the minimum of solar activity, the total cosmic-ray intensity is given by†

$$I_{CR} = \int_0^\infty I(E)\,dE \simeq 0.23 \text{ particle/cm}^2\text{-sr-sec.}$$

For the proton component, one has $I_p \simeq 0.20$.

If we computed the intensity of nucleons, one gets $I_{nucl} \simeq 0.38$. Then the corresponding values of N_{CR}, N_p, N_{nucl} are:

$$N_{CR} = 1.2 \times 10^{-10} \text{ particle/cm}^3,$$
$$N_p = 1.0 \times 10^{-10} \text{ proton/cm}^3,$$
$$N_{nucl} = 2.0 \times 10^{-10} \text{ nucleon/cm}^3. \tag{1.10}$$

Finally, the cosmic-ray energy density is equal to

$$w_{CR} = \int_0^\infty E_K N_{CR}(E)\,dE = \int_0^\infty \frac{4\pi}{v} E_K I(E)\,dE \simeq 0.9 \text{ eV/cm}^3,$$
$$E_K = E - Mc^2, \quad E = \frac{Mc^2}{\sqrt{1 - v^2/c^2}}. \tag{1.11}$$

The part due to the protons is equal to $w_p \simeq 0.65 \text{ eV/cm}^3$. Value (1.11) is characteristic because it is insensitive to the degree of solar activity and allows one to estimate the importance of the energy and of the pressure ($p_{CR} = w_{CR}/3$) of the cosmic rays in interstellar space. As a matter of fact, according to considerations connected with the application of Liouville's theorem and a general description of the motion of the cosmic rays in the Galaxy, and also taking into account their isotropy, one can conclude that value (1.11) is typical of a great part of the Galaxy. In later estimates we shall assume that in the Galaxy

$$w_{CR} \sim 1 \text{ eV/cm}^3 \sim 10^{-12} \text{ erg/cm}^3. \tag{1.12}$$

Note that for value (1.12), and with equipartition of the energy, i.e.

---

*Data on large atmospheric showers with $E \leq 5 \times 10^{15}$ also give some preliminary information on anisotropy ($\delta \sim 0.1\%$), the maximum intensity being in the direction of the galactic center.

†The integral $\int$ on $)\,dE$ converges, but its value is sensitive to the character of the spectrum in the range of low energies. For this reason, the quantities I_{CR} and $N_{CR} = 4\pi \int [I(E)/v]\,dE$ give only a rough value, whereas the energy w_{CR} depends only weakly on the contribution of low-energy particles. To prove what we have said, note that, according to the most recently available data (Ref. 11) which give the best estimates of that range, the total intensity of protons is $I_p = 0.2440 \pm 0.0160 \text{ proton/cm}^2\text{-sr-sec}$ and $I_{CR} = 0.28 \text{ particles/cm}^2\text{-sr-sec}$. The recalculated values are: $I_{CR} = 0.30$; $N_{CR} = 1.5 \times 10^{-10}$, and $w_{CR} = 0.8 \text{ eV/cm}^3$. The most recent calculations give the value $w_{cr} = .6 \text{ eV/cm}^3$ near the Earth at the solar minimum.

$$w_{CR} = \frac{H^2}{8\pi} = \rho \frac{u^2}{2},$$ (1.13)

the intensity of the magnetic field H is

$$H \sim 5 \times 10^{-6} \text{ oersted},$$ (1.14)

and for a gas density

$$\rho \sim 10^{-26} \text{ g/cm}^3 \text{ (concentration n} \sim 10^{-12}/\text{cm}^3)$$

the velocity of the gas is $u = 10^7$ cm/sec (such values are characteristic for the halo; in the disk one finds instead that $\rho \sim 10^{-24}$ g/cm^3 and $u \sim 10^6$ cm/sec).

The internal energy density of the gas is equal to $(3/2)nkT$ and $(3/2)nkT \sim w_{CR}$ for example if $n \simeq 1$ and $T \sim 10^4$ °K (HII regions of the disk) and if $n \sim 10^{-2}$, $T \sim 10^6$ °K (halo). As expression (1.6) (for the radius of curvature r of the particles) and corresponding estimates show, the field (1.14) is a strong field—in the sense that for a great part of the cosmic rays the radius r is very small, not only compared with the characteristic dimension of the Galaxy $L \simeq 3 \times 10^{22}$ cm, but also compared with the dimension of the gas nonuniformity in the disk (i.e. a distance $l \sim 10$ to 100 pc $\sim 10^{20}$ cm between gas clouds).

I.4. ELECTRONS AND POSITRONS. GAMMA AND X-RAYS

The existence of the electron-positron component of cosmic rays was well predicted quite some time ago (at least since 1953). As a matter of fact, the existence of this component follows directly from the fact that $\pi^\pm$ mesons are produced by cosmic rays in interstellar space; these mesons later decay by $\pi^\pm \to \mu^\pm \to e^\pm$. Also it is just the electron-positron component* which is responsible for the observed cosmic synchrotron (magnetobrems) radiation. However, electrons in the primary cosmic rays were only discovered in 1961, and even now information regarding the electron component is rather scarce.

Preliminary data, as of the present, is as follows (last data see in Ref. 30 and 34). In the energy region $1 < E < 10$ BeV, experimental data are consistent with the spectrum (see also Ref. 28).

$$I_e(E) = 40 \cdot E^{-2} \text{ electrons/m}^2\text{-sr-sec-BeV},$$ (1.15)

where E is measured in BeV; this spectrum is rather consistent with the data obtained for $I_e(>E = 4.5$ BeV), namely $I_e(>4.5$ BeV$) = 6.6 \pm 2.5$ electrons/m^2-sr-sec, which is $1.5 \pm 0.4\%$ of the proton intensity for $E > 4.5$ BeV.

For $E > 5 \div 10$ BeV, the exponent γ_e increases, so that $\gamma_e > 2$. However, for $E < 3 \div 5$ BeV, all data indicate that the exponent γ_e of the electron spectrum is significantly smaller than the exponent $\gamma \sim 2.5$-2.7 of the nuclear component spectrum. This state of affairs is a strong argument that the electrons in the region $E > 1$ BeV are not secondaries,† that is produced in the decays $\pi^\pm \to \mu^\pm \to e^\pm$, but are rather accelerated in the sources themselves.

---

*In most cases, with respect to synchrotron radiation for example, electrons and positrons are completely equivalent. Therefore, unless it could lead to some misunderstanding, we shall always refer to the electron-positron component as simply the electron component in the following.

†The secondary electron-positron component, in the region $E > 1$ BeV, should have the same spectrum as its generating component. In other words, in this case $\gamma_e \simeq \gamma \simeq 2.6$. When losses (particularly synchrotron and Compton losses) are taken into account, then for the secondary electrons $\gamma_e > \gamma \simeq 2.6$. On the other hand, the increase in the exponent of the observed electron spectrum for $E > 5 \div 10$ BeV is at most by 1; it is possible that this change in exponent is due to synchrotron and Compton losses (cf. Refs. 1 and 4).

Furthermore, that this is so, is also indicated by the observed absolute value of the intensity (this value is greater than what should be expected for secondary particles), and by radio-astronomical data (cf. Sec. 17 of Ref. 1; secondary electrons could not account for the observed radio-emission, neither for the intensity, nor the form of the spectrum). Finally, the problem of the nature of the electron component should be determined by measuring the ratio of the positron flux to the total flux [this ratio will be denoted by $e^+/(e^+ + e^-)$]. In the case of secondary electrons, we should have $e^+/(e^+ + e^-) > 0.5$. Preliminary results for $0.3 < E < 3$ BeV are (Ref. 3)

$$\frac{e^+}{e^+ + e^-} = 0.20 \pm 0.15. \tag{1.16}$$

For $E > 5$ BeV, according to available data $e^+/(e^+ + e^-) \lesssim 0.2$. Generally, the impression is that secondary electrons and positrons constitute only a small fraction of the electron component for $E > 1 \div 3$ BeV (according to our last information for $1 < E < 5$ BeV the ratio $e^+/(e^+ + e^-) = 0.05 \pm 0.03$): when the energy decreases, the fraction of secondary particles increases. Future study of the quantities $I_{e^-}(E)$ and $I_{e^+}(E)$ should be extremely interesting. Knowledge of the intensity and spectrum of the secondary particles will yield, among other things, information about the thickness $x_e \sim \rho c T_e$ of interstellar gas traversed by the cosmic rays, and (considering synchrotron and Compton losses) also information about the characteristic lifetime T_e of the secondary particle in the Galaxy, and about the average density ρ of the gas in the region traversed by the cosmic rays. With the electrons and positrons created in the decay $\pi^\pm \rightarrow \mu^\pm \rightarrow e^\pm$ neutrinos are obviously also created. But neutrinos can only be detected deep in the Earth. However, in this case, secondary neutrinos of cosmic origin are only a small fraction ($\sim 1\%$) of the neutrinos produced in the atmosphere. Thus the study of the neutrino component of the cosmic rays is very unlikely.*

On the other hand, the possibility of discovering high-energy gamma rays (which one could call the photon component of cosmic rays) is quite realistic. As regards softer photons (x rays), these have already been detected, and their study constitutes the newest and most promising branch of astronomy.

The processes leading to the creation of the x and γ cosmic rays have been discussed in Ref. 4. One can find in Ref. 3, 12 and 13 some observational results. The data concerning in particular the discrete sources, and especially the Crab nebula, although they are important for astrophysics in general, have only indirect relation to this course. Concerning the background continuum (quasi-isotropic) in the region 2 to 8 Å (i.e. energy between 1.5 and 6 KeV) the intensity of radiation is about 10 photons $cm^{-2} sr^{-1} sec^{-1}$. It is possible that this radiation is emission coming from a large number of galaxies and radiogalaxies but not resolved by the instruments. At the same time if one considers that the background continuum is due to diffusion of relativistic electrons by photons (in practice by thermal photons at 2.7 °K, ci. Sec. I.3) one can find the upper limit of the intensity of the cosmic-ray electron component in the Metagalaxy. For the same purpose one can use the data on the upper limit of the background continuum γ intensity given by

$$I_\gamma (> E = 40 \text{ MeV}) \leq (3.3 \pm 1.3)10^{-4} \text{ photons}/cm^2\text{-sr-sec.} \tag{1.17}$$

The estimation of the relativistic electron intensity in the Metagalaxy will be given in Sec. III.3 (see also Refs. 1 and 6).

---

*We disregard here the possibility, which we consider very improbable, that the intensity of high-energy neutrinos in the Universe is significantly greater than what could possibly be generated in cosmic-ray collisions.

II. COSMIC RAYS IN THE UNIVERSE

II.1. INTRODUCTION. SPECTRUM, PARTICLE DENSITY AND TOTAL ENERGY OF COSMIC RAYS IN SYNCHROTRON SOURCES

The study of cosmic rays on Earth permits, as we have seen, the acquisition of valuable information. But due to the isotropy of cosmic rays, it is impossible to obtain from this information about the galactic and extragalactic sources. One may thus see that cosmic ray astronomy could not be developed without the aid of radio-astronomy. Naturally we consider here the fact that the major part of cosmic radio emission is synchrotron radiation; from this fact and the data from radioastronomy one may obtain information concerning the intensity $I_e(E)$ of the electron component of cosmic rays in the Galaxy, in the envelopes of supernovae, in radiogalaxies, etc. We assume now a knowledge of the theory of synchrotron radiation and of the method of determining the intensity $I_e(E)$ from the information given by radioastronomy (cf. Refs 1 and 14 and Ref. 4); but, in the aim of simplification, we shall recall here the fundamental formulas and discuss the hypothesis used.

We assume the spectrum of the isotropic electron component to have the form

$$N_e(E)\,dE = K_e E^{-\gamma}dE, \quad N_e(E) = 4\pi I_e(E)/c, \qquad (2.1)$$

where the electrons are assumed to be ultrarelativistic as will be the case in what follows $(E \gg mc^2 = 10^5 \text{ eV})$. We will make further the assumption that the electrons are confined to a volume of dimension L along the line of sight and that the field in this volume is equal to H and is isotropic in average. Then the intensity of the synchrotron radiation in this direction is given by the formula:

$$\mathscr{I}_\nu = 1.35 \times 10^{-22} a(\gamma) L K_e H^{(\gamma+1)/2}\left(\frac{6.26 \times 10^{18}}{\nu}\right)^{(\gamma-1)/2} \text{ erg/cm}^2\text{-sr-sec-Hz}, \qquad (2.2)$$

where $a(\gamma)$ is of the order of 0.1 [for example, $a(2) = 0.103$ and $a(3) = 0.074$]; for a table of values of $a(\gamma)$, see Refs. 1, 4, and 14. As a result, for a power spectrum (2.1), the spectral index does not depend on γ (the definition of the spectral index α is $\mathscr{I}_\nu = \text{const } \nu^{-\alpha}$):

$$\alpha = (\gamma - 1)/2. \qquad (2.3)$$

The intensity $\mathscr{I}_\nu$ is proportional to $LK_e H^{(\gamma+1)/2}$ and, if one can evaluate L and H, one can determine K_e and the electron density $N_e(E)$. It is convenient to use here the formula

$$K_e = \frac{7.4 \times 10^{21} \mathscr{I}_\nu}{a(\gamma)LH}\left(\frac{\nu}{6.26 \times 10^{18}H}\right)^{(\gamma-1)/2}, \qquad (2.4)$$

where, as in the other formulae, K_e is measured in $(\text{ergs})^{\gamma-1}$ cm^{-3}, L in cm, H in oersteds, ν in Hz, and $\mathscr{I}_\nu$ in ergs (cm^2 sr-sec-Hz)$^{-1}$.

We remark in passing that the effective temperature of the radiation is given by the formula

$$T_{\text{eff}} = \frac{c^2}{2k\nu^2}\,\mathscr{I}_\nu, \qquad (2.5)$$

where $k = 1.38 \times 10^{-16}$ erg/degree.

If the electron distribution along the line of sight is not homogeneous, but if H is constant, the product $K_e L$ in (2.2) may be replaced by $\int_L K_e dL$.

If one observes the discrete radio sources one does not measure $\mathscr{I}_\nu$ but the flux density $\Phi_\nu = \int \mathscr{I}_\nu d\Omega$, where the integration is made over the solid angle substended by the source. For a small enough source (more exactly, for a source whose largest dimension L is much smaller than the distance to the observer, R), and if K_e and H are assumed to be constant in the source, one obtains from (2.2):

$$\Phi_\nu = 1.35 \times 10^{-22} a(\gamma) \frac{K_e V H^{(\gamma+1)/2}}{R^2} \left(\frac{6.26 \times 10^{18}}{\nu}\right)^{(\gamma-1)/2} ,$$

$$K_e = \frac{7.4 \times 10^{21} R^2 \phi_\nu}{a(\gamma) H V} \left(\frac{\nu}{6.26 \times 10^{18} H}\right)^{(\gamma-1)/2} \tag{2.6}$$

where V is the volume of the source (for a spherical source with radius r = L/2 one has naturally $V = \pi L^3/6$).

From (2.6) one sees that the number of relativistic electrons is given by

$$N_t(> E_1) = V \int_{E_1}^{E_2} K_e E^{-\gamma} dE \simeq \frac{7.4 \times 10^{21}}{(\gamma-1) \times a(\gamma)} \frac{R^2 \Phi_\nu}{H} \left[\frac{y_1(\gamma)\nu}{\nu_1}\right]^{(\gamma-1)/2} , \tag{2.7}$$

where E_1 and E_2 are the limits of the electron energy interval, inside of which the spectrum (2.1) is valid, and the frequencies ν_1 and ν_2 are related to E_1 and E_2 by the equations

$$E_1 = mc^2 \left[\frac{4\pi mc\nu_1}{3eHy_1(\gamma)}\right]^{1/2} \simeq 2.5 \times 10^2 \left[\frac{\nu_1}{y_1(\gamma)H}\right]^{1/2} \text{ eV},$$

$$E_2 \simeq 2.5 \times 10^2 \left[\frac{\nu_2}{y_2(\gamma)H}\right]^{1/2} \text{ eV}, \tag{2.8}$$

$$y_1(1) = 0.80; \quad y_1(1.5) = 1.3; \quad y_1(2) = 1.8; \quad y_1(2.5) = 2.2; \quad y_1(3) = 2.7;$$

$$y_2(1) = 4.5 \times 10^{-4}; \quad y_2(1.5) = 0.011; \quad y_2(2) = 0.032; \quad y_2(2.5) = 0.10;$$

$$y_2(3) = 0.18.$$

In (2.7) we have assumed that ν_1 is much less than ν_2 and that γ is greater than 1; then the number of electrons is determined by the lower limit of the frequency interval. (A more general formula is given in Refs. 1 and 14.) The total energy of electrons in the source radiating in the interval $\nu_1 \le \nu \le \nu_2$ is given by the formulae

$$W_e = V \int_{E_1}^{E_2} K_e E^{-\gamma+1} dE = A(\gamma,\nu) \frac{R^2 \Phi_\nu}{H^{3/2}} ,$$

$$A(\gamma,\nu) = \frac{2.96 \times 10^{12}}{(\gamma-2)a(\gamma)} \nu^{1/2} \left[\frac{y_1(\gamma)\nu}{\nu_1}\right]^{(\gamma-2)/2} \left\{1 - \left[\frac{y_2(\gamma)\nu_1}{y_1(\gamma)\nu_2}\right]^{(\gamma-2)/2}\right\} \text{ for } \gamma > 2,$$

$$A(\gamma,\nu) = 1.44 \times 10^{13} \nu^{1/2} \ln \left[\frac{y_1(\gamma) \times \nu_2}{y_2(\gamma) \times \nu_1}\right] \text{ for } \gamma = 2,$$

$$A(\gamma,\nu) = \frac{2.96 \times 10^{12}}{(2-\gamma)a(\gamma)} \nu^{1/2} \left[\frac{y_1(\gamma)\nu}{\nu_2}\right]^{(\gamma-2)/2} \left\{1 - \left[\frac{y_2(\gamma)\nu_1}{y_1(\gamma)\nu_2}\right]^{(2-\gamma)/2}\right\} \text{ for } \frac{1}{3} < \gamma < 2. \tag{2.9}$$

We have assumed here that the frequency interval is sufficiently large, so that $\nu_2/\nu_1 \gtrsim y_1(\gamma)/y_2(\gamma)$; moreover the formula giving $A(\gamma, \nu)$ in fact is valid only when $\gamma > 1.5$ (or $\alpha > 0.25$). If the frequency interval is small, or if γ is less than 1.5, then one can obtain only a rough estimation of the electron energy. We assume in this case that every electron of energy E radiates on the frequency

$$\nu_m = 0.07 \frac{eH_\perp}{mc}\left(\frac{E}{mc^2}\right)^2 = 1.2 \times 10^6 H_\perp \left(\frac{E}{mc^2}\right)^2 = 1.8 \times 10^{18} H_\perp [E(\text{erg})]^2 =$$

$$4.6 \times 10^{-6} H_\perp [E(\text{eV})]^2. \tag{2.10}$$

Further it is then necessary in (2.8) to set $y_2(\gamma) \simeq y_1(\gamma) \simeq 0.24$. For the frequency ν_m the spectral power of radiation of one electron is

$$p_m = p(\nu_m) = 1.6\frac{e^3 H_\perp}{mc^2} = 2.16 \times 10^{-22} H_\perp \text{ erg/(sec-Hz)} \tag{2.11}$$

and for a discrete source containing $NV = N_t$ electrons the radiation flux is

$$\Phi_\nu = \frac{p_m NV}{4\pi R^2}, \quad N_t = \frac{4\pi R^2 \Phi_\nu}{p_m} \simeq 5 \times 10^{22} \frac{R^2 \Phi_\nu}{H}, \tag{2.12}$$

where one has put, for comparison with (2.7), $H_\perp^2 = (2/3)H^2$ ($H_\perp$ is the component of the field $\vec{H}$ perpendicular to the line of sight, this relation corresponds to the case where the magnetic field H is isotropic in average).

The energy in the source is [still assuming that $H_\perp^2 = (2/3)H^2$]

$$W_e = N_t E \simeq 5 \times 10^{13} \nu^{1/2} \frac{R^2 \Phi_\nu}{H^3/2} \text{ erg}, \tag{2.13}$$

because in the condition (2.10) we have

$$E = 7.5 \times 10^{-10} \left(\frac{\nu}{H_\perp}\right)^{1/2} \text{erg} = 4.7 \times 10^2 \left(\frac{\nu}{H_\perp}\right)^{1/2} \text{ eV}. \tag{2.14}$$

The approximations (2.12) and (2.13) give minimum values for N_t and W_e because one considers that the electrons essentially emit at the frequency ν.

Equations (2.9) and (2.13) permit an estimate of the electron energy in the source when one knows the radiation flux Φ_ν, the distance to the source R, and the average field H in the source. The first two quantities can be determined, but the field H cannot be determined in an independent manner.* It is thus necessary to formulate a complementary hypothesis. One important hypothesis is that the magnetic energy in the source $W_H = (H^2/8\pi)V$ is proportional to the total energy of the cosmic rays:

$$W_H = \frac{H^2}{8\pi} V = \kappa_H W_{CR}. \tag{2.15}$$

The result of (2.9) and (2.15) is that the energy W_{CR} increases when H decreases, (see also (2.17)) whereas the energy W_H decreases. One therefore sees immediately that the total energy $W_t = W_{CR} + W_H$ is at a minimum or near a minimum for $W_H \sim W_{CR}$, i.e. for $\kappa_H \sim 1$ (at minimum, one has $\kappa_H = 3/4$ because $W_t = c_1 H^2 + c_2 H^{-3/2}$). The assumption $\kappa_H \sim 1$ is reasonable also when one considers the dynamics: when W_H is much smaller than W_{CR} (i.e. $\kappa_H \ll 1$) the field cannot retain the cosmic rays inside the source for a long time; if $\kappa_H \gg 1$, the synchrotron radiation is very intense and the electrons lose their energy very rapidly.

Unfortunately, despite the hypothesis of (2.15) and the assumption

$$\kappa_H \sim 1 \tag{2.16}$$

it is still necessary to relate the cosmic-ray energy W_{CR} and the energy of the electrons W_e. With this in mind we put:

$$W_{CR} = \kappa_e W_e. \tag{2.17}$$

---

*In principle this determination is possible: for example by measuring the Zeeman splitting of hydrogen lines (either the 21-cm line or lines of excited hydrogen). This possibility is however not real for the supernovae and radiogalaxies.

Usually it is natural to put, in agreement with cosmic-ray measurements made near the Earth, that

$$\kappa_e \sim 100. \tag{2.18}$$

But, in such estimating κ_e one has less information than in the case of κ_H. If the acceleration mechanism in the source was well understood, one would be able to evaluate κ_e but in fact it is still impossible. One may only remark that in the mechanisms discussed usually κ_e is greater than or equal to 2; in addition, even if in the course of acceleration κ_e is ~ 2, the electrons lose their energy very rapidly by synchrotron radiation and Compton scattering. The parameter κ_e increases as a result. For this reason the assumption that $\kappa_e \gg 1$, and, more precisely estimation (2.18) provide, generally, a reasonable estimate. From conditions (2.15) and (2.17) and from (2.9), one obtains

$$W_{CR} = \kappa_e W_e = 0.19 \times \kappa_H^{-3/7} \left[\kappa_e A(\gamma, \nu) \Phi_\nu R^2 \right]^{4/7} (R\varphi)^{9/7},$$

$$W_H = V \frac{H^2}{8\pi} = \kappa_H W_{CR}, \quad H = \left[48 \kappa_H \kappa_e A(\gamma, \nu) \frac{\Phi_\nu}{R\varphi^3} \right]^{2/7}, \tag{2.19}$$

where the volume of the source V is $(\pi/6)L^3$ and the angular diameter is $\varphi = L/R$.

With the aid of the several assumptions given here, and from radioastronomical data, one is there able to determine the electron spectrum and the total cosmic-ray energy in the sources. Utilization of the theory of synchrotron radiation permits one to obtain also further information. For example, measurement of polarization supplies data on the structure of the field and the plasma density along the line of sight (cf. Refs. 1, 4, and 14). But we limit ourselves here to a brief discussion of a different problem. Let us assume that all the electrons have been accelerated at one instant $t = 0$ and that subsequently they lose their energy only by synchrotron radiation.*

An electron which had energy E_0 at time $t = 0$, has now an energy given by

$$E = \frac{E_0}{1 + \beta E_0 t}, \quad \beta = \frac{2e^4}{3m^4 c^7} H_\perp^2 = 1.95 \times 10^{-9} \frac{H_\perp^2}{mc^2} \text{ erg}^{-1} \text{ sec}^{-1} \tag{2.20}$$

One sees clearly that the energy is reduced by a factor of 2 after a time given by

$$T_m = \frac{1}{\beta E_0} = \frac{5.1 \times 10^8}{H_\perp^2} \frac{mc^2}{E} \text{ sec}, \tag{2.21}$$

and one has the relation

$$E(t) \le E_m(t) = \frac{1}{\beta t} = \frac{2.6 \times 10^{14}}{H_\perp^2 t} \text{ eV}. \tag{2.22}$$

One thus would see in the radiation spectrum a "break" at frequencies $\nu \sim \nu_m(E_m(t))$ with [cf. (2.10)]

$$\nu_m(E_m(t)) = \frac{3.1 \times 10^{23}}{H_\perp^3 t^2} \text{ Hz}. \tag{2.23}$$

We also see changes in the spectrum from other causes (starting from the source at time $t = 0$, the electrons are then leaving the system by diffusion, etc.) One can also consider changes in the radiation flux Φ_ν originating in the source

---

*Under certain conditions, one can obtain a similar result for Compton losses by replacing $H^2/8\pi$ by $\overline{w}_{ph}$ (cf. Refs. 1, 4, and 14).

TABLE II. Galactic and Extragalactic Radio Sources.

Source	Angular dimension φ	Distance R	Spectral index α	Flux for $\nu = 10^8$ Hz (W/m^2 sec Hz)	Radio luminosity L_r in the interval $10^7 < \nu < 10^{10}$ Hz (W/sec)	Magnetic field (oersted)	Total cosmic-ray energy (ergs) W_{CR}
Cassiopeia A	4'	3400 pc	0.8	1.9×10^{-22}	2.5×10^{35}	1.3×10^{-3}	6.6×10^{49}
Taurus A (Crab Nebula, SN 1054)	5'	1700 pc	0.25 / 0.35 (optics)	1.7×10^{-23} / —	2.4×10^{34} / 3.1×10^{37} (for $\nu_2 = 10^{15}$ Hz)	9×10^{-4} / 1.2×10^{-3}	8.0×10^{48} / 1.5×10^{49}
Tycho Brahé, SN 1572	5.4'	3000 pc (in fact $\gtrsim$ 3000 pc)	0.6	2.4×10^{-24}	3.8×10^{33}	3.0×10^{-4}	5.5×10^{48}
Kepler SN 1604	2'	6000 pc	0.6	8×10^{-25}	5.0×10^{33}	4.0×10^{-4}	4.6×10^{48}
Our Galaxy	—	—	0.8	—	4.4×10^{38}	6.0×10^{-6}	3.0×10^{56}
Galaxy M31 (Andromeda)	200'	0.73 Mpc	0.5	2.7×10^{-24}	3.3×10^{38}	2.9×10^{-6}	4.0×10^{56}
Cygnus A	$2 \times 38''$	220 Mpc	0.75	1.3×10^{-22}	7×10^{44}	1.9×10^{-4}	3.0×10^{60}
Centaurus A	$2 \times 180'$ (extended source)	3.8 Mpc	0.77	7.6×10^{-23}	1.3×10^{41}	4×10^{-6}	1.7×10^{59}
	$4.5'$ (central source)			1.8×10^{-23}	3×10^{40}	8×10^{-5}	5.0×10^{56}
Virgo A	10' (extended source)	11 Mpc	0.8	6×10^{-24}	8×10^{40}	2.2×10^{-5}	1.0×10^{58}
	$2 \times 23''$ (central source)	11 Mpc	0.8	9×10^{-24}	1.2×10^{41}	3.3×10^{-4}	2.5×10^{56}

(for example in the envelopes of supernovae) as a result of its expansion. Reabsorption and the influence of the surrounding plasma also determine changes in the spectrum (cf. Refs. 1, 4, and 14). So we see that data from radioastronomy and, of course, information about optical and x radiation from synchrotron process really allow one to make conclusions about cosmic rays far from the Earth.

II.2. SOME INFORMATION ABOUT COSMIC RAYS FAR AWAY FROM THE EARTH

Radio observations have established beyond doubt that supernova envelopes contain a large quantity of relativistic electrons. One measures the radiation flux Φ_ν; if this flux covers a large frequency interval and if one knows the distance R to the source, one is able to determine the emitted power, in other words, the radio luminosity of the source:

$$L_r = 4\pi R^2 \int_{\nu_1}^{\nu_2} \Phi_\nu \, d\nu. \tag{2.24}$$

For purposes of normalization one chooses very often the values $\nu_1 = 10^7$ cps and $\nu_2 = 10^{10}$ cps. In Table II are presented the values of L_r determined for a series of sources, as well as other parameters. The field H and the cosmic-ray energy W_{CR} have been calculated by using Eq. (2.19), with $\kappa_H = 1$ and $\kappa_e = 100$.* The data used in Table II are the same as in Ref. 1; only the distances to type I supernovae (1054, 1572, and 1604) have been changed by recent data, and in fact they are still under discussion. In recent literature one can find values of R, α, and Φ_ν which are different from those given here. For example, if one uses the most probable value for the Hubble constant, H = 100 km/(sec Mpc), the distance to Cygnus A becomes 165 Mpc, as contrasted with 220 Mpc when H = 75 km/(sec Mpc) is used. But we have not changed these values, for our goal is only an estimation of the field H and the energy W_{CR}. In more exact calculations, one cannot assume that the spectral index is a constant, hence equations (2.19) can be used only for estimates.

From Table II and from more complete data (cf. Ref. 1) one sees that in the most powerful radiogalaxies (Cygnus A and 3C 295) the energy W_{CR} is of the order 3×10^{60} ergs. The total energy necessary for the formation of a radiogalaxy is still larger. Firstly, the minimum total energy is given by $W_t = W_{CR} + W_H \simeq 6 \times 10^{60}$ ergs; secondly a portion of the cosmic rays has left the system, carrying with them part of the energy, and some of the energy is contained in the form of kinetic and internal energy of the gas. One thus arrives at a total energy $W_t \simeq 10^{61}$ to 10^{62} ergs, which corresponds to 10^7 to 10^8 $M_\odot c^2$. (The mass of the sun $M_\odot$ is 2×10^{33} g.) One probably has here a situation which eventually results in a great explosion, the most powerful is the Universe, except connected with the evolution of the Metagalaxy.

For the majority of the radiogalaxies, W_t is less than the value for Cygnus A. But for the relatively strong sources Centaurus A and Virgo A the energy W_{CR} is of the order of 10^{58} to 10^{59} and W_t is of the order of 10^{59} to 10^{60} ergs. For "normal galaxies" (eg. the galaxy M31) a typical value is $W_{CR} \sim 1$ to 3×10^{56} ergs.

The total energy of cosmic rays in supernova envelopes is $W_{CR} \sim 10^{49}$ ergs, and if one considers the emission of particles and the resulting losses, one can infer that the initial explosion has liberated $W_{CR,0} = 10^{50}$ ergs. The upper limit of $W_{CR,0}$ is determined by the energy liberated in the supernova explosion, which can attain a value of 10^{51} to 10^{52} ergs. However, not only the acceleration of electrons but also of protons and nuclei is an important question in the case of supernovae. If for example one takes $\kappa_e = 1$ for the Crab Nebula (only electrons are accelerated), one finds with the same other assumptions as above that the energy $W_{CR} = W_e$ is $\kappa_e^{4/7} \sim 10$ times less [cf. (2.19)]; one has assumed before that $\kappa_e = 100$ and as a result $W_{CR} \sim 10^{48}$ ergs.

---

*For our Galaxy the calculations are made by replacing in (2.19) $R^2\Phi_\nu$ by $V\mathscr{I}_\nu/L$, where one assumes that V is 3×10^{68} cm^3, that the diameter is L = 10000 pc, and $\mathscr{I}_\nu$ corresponds to the effective temperature T_{eff} = 25 °K at ν = 400 cps.

More important is that if in a supernova explosion only electrons are accelerated, the supernovae cannot be sources of the cosmic-ray protons and nuclei observed on the Earth. However as already has been said, it is probable that in the majority of the explosions (the Crab nebula may be an exception) κ_e is much greater than one.

As far as the origin of cosmic rays observed near the Earth is concerned, we are interested here in discrete galactic sources (primarily in the envelopes of supernovae), in the Galaxy itself, and finally in sources contained in the Local Group and in intergalactic space. We have already discussed supernova envelopes. Radio radiation from intergalactic space has not yet been detected and is probably very weak. This is perhaps due to the weakness of the intergalactic field ($H_{ing} \lesssim 10^{-8}$ oersteds) and to a low density of the electron component of intergalactic cosmic rays. This last conclusion comes from the data of x and γ-ray astronomy (cf. Sec. III.3). As far as the intergalactic nucleon and proton C. R. components are concerned one is not able to make direct conclusions and has to analyse the problem indirectly (cf. Sec. III.3).

As for general galactic radiation,* one has at his disposal a large amount of information, but unfortunately it is inexact and quite non-unique. It is a curious fact that several years ago the situation appeared to be clearer than it is now. It was thought possible to separate galactic radiation into radiation coming from the halo and radiation coming from the disk (cf. Fig. 2, and for the details see Ref. 1, Sec. 5 and Ref. 15). Now it is beginning to be thought that the disk is not a continuous source but is constituted of a number of radiating regions. Most important is that one can admit doubt about the existence of the radio halo, or at least of a very large radio halo (R ~ 10 to 15 kpc). Establishing the existence of a radio halo from earthbound observations is a difficult task, because it is not possible to distinguish between the quasi-isotropic radiation of a halo and the continuum of the Metagalaxy. One is not able to solve the problem of the structural composition of the halo because one does not have direct methods for determining the distance of sources in the radio sky. Naturally it is necessary to continue research in this area; only observations can resolve the situation. In any case one cannot now see any reason for the nonexistence of a well defined, large halo. Radio-astronomical data on the Galaxy itself (cf. Refs. 15 and 16) and on other special galaxies with a possible radio halo (cf. Ref. 17) verify the above conclusion. We only remark here that the ideas concerning the existence of a halo have been introduced (cf. Ref. 18) in order to explain the trapping of cosmic rays in the Galaxy. These considerations and a general analysis of the character of cosmic-ray motion in the Galaxy (Ref. 1 and 19) support the hypothesis of the existence of a halo. We shall discuss the details in Sec. III.2 when we discuss galactic models of the origin of cosmic rays (see also Ref. 28).

According to the latest data available, (Ref. 16) radiation from the galactic halo is well approximated by a power spectrum $\mathcal{I}_\nu = \text{const } \nu^{-\alpha}$ with $\alpha = 0.43 \pm 0.03$ in the large frequency range from 10 to 400 MHz (i.e. $\lambda = 75$ cm to 30 cm). One then has $\gamma = 2\alpha + 1 = 1.86 \pm 0.06$, and we speak about the electrons which have an energy between 10^9 and 5×10^9 eV. (This corresponds to a field $H = 2 \times 10^{-6}$ oersted [cf. (2.14)]. The value $\gamma = 1.86 \pm 0.06$ is not contradicted by the data on electrons seen near the Earth (cf. Sec. I.4). One is able to say the same thing about the intensity of the electron component, at least to within the observational errors. For example, in Ref. 20 one assumes that in the direction of the galactic pole the halo has a dimension of 12 kpc and that the effective temperature of the radiation at 178 MHz is $T_{eff} = 110$ °K. At the same time this radio radiation of the halo may be explained if one considers that the electrons in the halo with $H \sim (2 \text{ to } 3) \times 10^{-6}$ Oe

---

*We consider general galactic radiation to be radiation coming not from discrete sources but from different regions of the Galaxy, or, in other words, radiation which continuously varies with the galactic coordinates.

have the same intensity as observed near the Earth.* In the future one will probably
be able to make a qualitative comparison of the different data, and one may make an
estimate of the average field in the halo starting from the values of the electron intensity and the radio radiation intensity. At present we can say only that the repre-
sentation is self-consistent, i.e. the estimation obtained for the magnetic field
(H ~ (2 to 3) × 10^{-6} Oe) is completely reasonable. One yet sees in these facts an
extra agrument in favor of the existence of a halo and for a galactic theory discussed later on the origin of cosmic rays observed near the Earth.

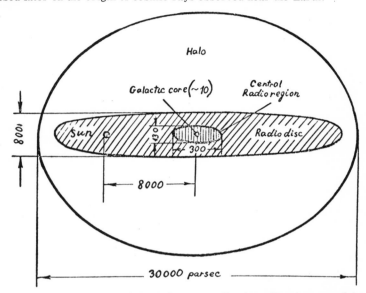

FIG. 2. Structure of the Galaxy according to radioastronomy data
in the metric waveband. The drawing is schematic and the dimensions are given in parsecs.

III. THE ORIGIN OF COSMIC RAYS

III.1. WHAT IS KNOWN AND UNKNOWN IN COSMIC-RAY ASTROPHYSICS? CHOICE OF MODELS.

The great role played by cosmic rays in the Universe is due to their large
energy density w_{CR}, to the pressure that they exert $p_{CR} = w_{CR}/3$ and to the universality of the acceleration mechanism (by universality we mean that this process
takes place practically everywhere). We have already insisted on these facts at the
beginning of the course and we have demonstrated this in Secs. I and II. Besides
the general considerations, one can obtain a series of concrete conclusions on the
cosmic rays in the Universe and in the vicinity of the Earth. It is not useful to re-
evoke here these conclusions; on the contrary we wish to emphasize the existence
of numerous unsolved problems and show in this way the prospects of future research.

*In Ref. 1, Sec. 17 more detailed calculations have been made but higher values
of radio intensity were used. If one considers this fact (i.e. if one uses T_{eff} = 110 °K
for ν = 178 MHz) the calculations of Ref. 1 are in agreement with observational
data. For information on the Galaxy and its radioemmision see Ref. 31.

TABLE III. Origin of Cosmic Rays Observed Near the Earth (Models Discussed).

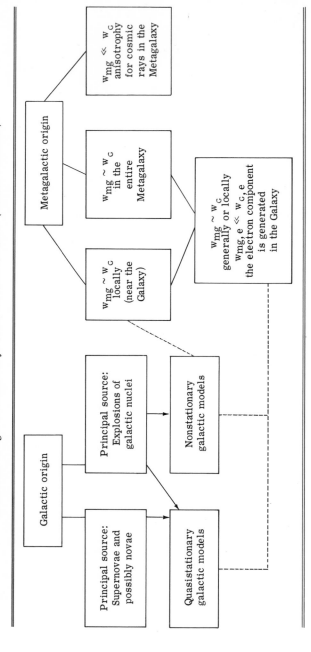

The most important problems in cosmic-ray astrophysics which have to be solved or made clearer are the following:

(1) The fundamental problem is the choice of a model (theory) for the cosmic rays observed near the Earth. Are the majority of these cosmic rays generated in the Galaxy or in the Metagalaxy? This is a question that we will meet further in a detailed manner.

(2) What is the acceleration mechanism of cosmic rays in their sources? In what way is the injection produced and the following acceleration? Is there preferential acceleration for the heavy nuclei and in what proportion compared with the nucleons are the electrons accelerated?

(3) A detailed analysis of the characteristics of the motion of the cosmic rays in interstellar and intergalactic space is necessary. It is not yet evident that in this problem one can apply the diffusion approximation (cf. Ref. 1, and Sec. III.2). And we do not know definitely the role played by plasma effects.

(4) One can, in particular, mention two special problems, not yet clear: a) the problem of cosmic rays of low energy ($\epsilon \leq 1\,\mathrm{BeV/nucleon}$) in interstellar space and the problem of the existence of a maximum in the energy spectrum of these cosmic rays and b) the problem of the cosmic-ray anisotropy (it is first of all an experimental question; this anistropy has to be measured, and then we have to establish definitely the reason for which we observe a high degree of isotropy on the Earth).

(5) As we have partly seen in Secs. I and II a number of other problems arise which must be the object of future theoretical and experimental research. Those are the problems of the chemical composition and the transformation of the cosmic rays, the problem of the electron-positron component, the characteristics of cosmic rays of very high energy, greater than $10^{15}\,\mathrm{eV}$ (i.e. the energy spectrum, the anisotropy, and the composition), the problem of the galactic halo, the problem of the compact source in the Crab and the nature of the x rays from this object, and finally the fundamental questions of the structure and the nature of radiogalaxies and quasars.

We will not consider in the following the problem of the acceleration mechanisms (cf. Ref. 1; the progress in this field was found in Ref. 21).

The ensemble of the problems given in (5) needs new experimental observations and we will not dwell on it. We will concentrate our attention on problem 1: on the choice of the model and on the discussion of the galactic and metagalactic theory of the origin of cosmic rays (Sec. III.2 and III.3). We will also discuss the problems connected with plasma effects (cf. Sec. III.4) because they are related with cosmic-ray astrophysics and one began to be interested in them a short time ago (cf. Refs. 6 and 10).

Problem (1) now reduces to a choice between several possibilities illustrated in Table III. A more detailed discussion of the models will be made later. Here we emphasize the fact that it is difficult to make a choice of the models for three reasons: the lack of direct information about the proton-nucleon component in intergalactic space; the lack of accurate knowledge of the structure of the galactic halo and of the configuration of the magnetic field in the halo and on the boundaries between the halo and the intergalactic space; the lack of direct data on the proton-nucleon component in the envelopes of supernovae.

About the electronic component of cosmic rays the situation is different because its galactic origin now seems to be certain. As it will be shown later, the theory of the galactic origin of cosmic rays also is much better founded than the theories of the metagalactic origin.

III.2. THE GALACTIC MODEL OF THE ORIGIN OF COSMIC RAYS

The most probable quasistationary galactic model of the origin of cosmic rays can be reduced to the following. The cosmic rays observed near the Earth (except the particles of low energy of solar origin and the particles of very high energy which probably come from the Metagalaxy) are produced in the Galaxy most of the time by the explosions of supernovae. One can also consider as supplementary "reserve" sources the explosions of novae and the "small" explosions of galactic nuclei (cf. later). To avoid using sources that are too powerful and for other reasons (connected to the isotropy of the cosmic rays, to the data on the chemical composition and to other information) one has to introduce the existence of a halo filled with cosmic rays which do not leave the halo too quickly. In practice the lifetime of cosmic rays in the halo is in agreement with the estimate (1.5) that it is of the order of $T \simeq (2 \text{ to } 3) \times 10^8$ years. By using a diffusion theory of the cosmic rays in the halo with an effective diffusion coefficient $D = lv/3$, one finds for the trapped time in the halo

$$T_h \simeq \frac{R^2}{2D} \simeq \frac{R^2}{lv}, \tag{3.1}$$

where l is the effective mean free path and $v \sim 10^{10}$ cm/sec is the speed of the translational movement between the collision with the inhomogeneities of the magnetic field (for more details, see Ref. 1). For $R \sim 3 \times 10^{22}$ cm and $T_h \sim 10^{16}$ sec, the coefficient D is of the order of 10^{29} cm/sec^2 and l is of the order of 10^{19} cm $\simeq 10$ pc. If one considers the time needed to leave the galactic radio disk whose half thickness $R_d \sim 300$ pc $\sim 10^{21}$ cm for $l = 10$ pc, one finds $T_d \sim 10^{13}$ sec $\sim 3 \times 10^5$ years.

The volume of the halo is $V \sim (4\pi/3)R^3 \sim 10^{68}$ cm^3. By using the considerations already presented (Liouville's theorem, isotropy of the cosmic rays) the energy density of the cosmic rays in the halo is nearly constant in the context of the model discussed and is $w_G \sim 10^{-12}$ erg/cm^3 [cf. (1.12)]. The index CR of the formula (1.12) has been removed and we will indicate instead, the place where the energy w is given—in the preceding case it is the Galaxy, represented by the index G.

One can deduce that the total energy of the cosmic rays in the Galaxy is

$$W_{CR} \sim w_G V \sim 10^{56} \text{ erg}. \tag{3.2}$$

The energy density of the electron component is:

$$W_e \sim (1-3) 10^{54} \text{ erg}. \tag{3.3}$$

If we take as the lifetime of the cosmic rays $T = 10^{16}$ sec, we obtain for a quasistationary state, a total power of the cosmic-ray sources

$$U_{CR} \sim \frac{W_{CR}}{T} \sim 10^{40} \text{ erg/sec}, \quad U_e \sim \frac{W_e}{T_e} \sim 3 \times 10^{38} \text{ erg/sec}, \tag{3.4}$$

where T_e is the lifetime of the electrons, which is slightly less than T because of the synchrotron and Compton losses (we note however that in this case one also gets $T_e \sim T \sim 10^{16}$ sec).*

*The energy of one electron decreases by a factor of 2 during the time $T_{m,c} \sim \dfrac{3 \times 10^7}{H^2/8\pi + w_{ph}} \dfrac{mc^2}{E}$ where w_{ph} is the energy density of radiation [cf. (2.21) and Ref. 4]; in the Galaxy, one has $H^2/8\pi + w_{ph} \sim 10^{-12}$ erg/cm^3 and $T_{m,c} \sim 3 \times 10^{16}$ sec for $E \sim 5 \times 10^8$ eV, and $T_{m,c} \sim 3 \times 10^{15}$ sec for $E \sim 5 \times 10^9$ eV.

Following the lack of exact data the value of U_{CR} and U_e can differ from the estimations (3.4) by one order of magnitude. However, if the model in question is valid, it is not very probable that the result differs greatly from the value that has been found.

The powers (3.4) are very high. For example, the Sun emits in cosmic rays an average of 10^{24} ergs/sec. It is for this reason that one would obtain only 10^{35} to 10^{36} ergs/sec if all the stars of the Galaxy emitted cosmic rays at the same rate. There are naturally stars that are more active than the Sun but they exist only in relatively small numbers. Considerations of this sort show that the problem of production of cosmic rays in the Galaxy is not easy (it is difficult to satisfy the energy requirements). On the other hand the possible role played by supernovae explosions appears clearly. In fact, during the explosion of a supernova, 10^{50}-10^{51} ergs are released and even sometimes 10^{52} ergs. It is for this reason that even before the discovery of the radio emission of supernovae one thought of the probable role of the supernovae as cosmic-ray sources. It is now firmly established that during the explosion of supernovae there is injection of a great quantity of relativistic electrons which are responsible for the synchrotron radiation of the supernovae envelopes. Even if one admits that there is no injection of a proton component, the electronic component has at least a total energy of 10^{48} ergs in average per supernovae. If one considers that a fraction of the particles has escaped from the envelopes in the beginning and if one takes into account the fact that the injection of the electronic component is also at present observed in the Crab Nebula then one gets an energy $W_{SN,e} \sim 10^{48}$-40^{49} ergs emitted per the explosion of the supernovae by supposing that the major part of the injected particles are electrons. In the Galaxy there is on the average one explosion every 100-300 years* and the power of the electronic component injection is given by

$$U_{SN,e} \sim \frac{W_{SN,e}}{T_{SN}} \sim \frac{10^{48} - 10^{49}}{3 \times 10^9 - 10^{10}} \sim 10^{38} - 3 \times 10^{39} \text{ erg/sec.} \qquad (3.5)$$

If one agrees that during the explosion of a supernovae, the proton component is also injected, in the ratio $\kappa_e \sim 100$ [cf. (2.18)], then the energy W_{SN} is of the order of 10^{49} to 10^{50} ergs and the injection power is

$$U_{SN} \sim \frac{W_{SN}}{T_{SN}} \sim \frac{10^{49} - 10^{50}}{3 \times 10^9 - 10^{10}} \sim 10^{39} - 3 \times 10^{40} \text{ erg/sec.} \qquad (3.6)$$

The estimates (3.4), (3.5), and (3.6) imply that the supernovae explosions can guarantee an adequate rate of generation of cosmic rays in the Galaxy. At the same time, it is not absolutely clear from these estimates that the supernovae really supply the proton-nuclei component, i.e. that they are the main sources of injection of the cosmic rays. It is also not quite clear that supernovae inject electronic component because the energy estimates are insufficient. In this connection one can note that novae and particularly galactic nuclei explosions can be alternative or supplementary sources of cosmic rays. The existence of such explosions in the nucleus of the Galaxy is not yet proved but one cannot reject this possibility [the explosions could have taken place $(1 \text{ to } 5) \times 10^7$ years ago]. Perhaps now we can say that such "weak" explosions are even probable. If the galactic nuclei explosions exist and are "weak" with liberation of a cosmic-ray energy $W_{nucl} \sim (1 \text{ to } 3) \times 10^{55}$ ergs, they could play the same role as supernovae explosions (for $W_{nucl} \sim 10^{55}$ ergs and one explosion each 3×10^7 years the injection power is $U_{nucl} \sim 10^{40}$ ergs/sec). In other words the quasistationary picture is respected because the 10 to 20% intensity fluctuations of the cosmic rays having taken place $(1 \text{ to } 3) \times 10^7$ years before would not be observable. One can equally imagine one "big" explosion during which would have been created the majority of the cosmic rays observed now and, possibly, the present halo. In this case one would be considering a galactic model that was nonstationary

---

*The best estimate now is one explosion every 50 years in average (see Sky and Telescope 33, N1, 3 (1967)). Last figure is even 30 years.

(cf. Table III). Its principal characteristic, as far as the cosmic-ray problem is concerned, would be a strong dependence (with a scale of 10^8 years) of the intensity of all the components on the time. While the hypothesis of weak galactic explosions is more or less acceptable (although not actually established), the hypothesis of one great galactic explosion has no serious basis. In fact, all arguments are against weights this possibility and against a nonstationary model of the origin of the cosmic rays. Firstly one great explosion implies $W_{nucl} \sim 10^{58}$ to 10^{59} ergs, and that our Galaxy was a radiogalaxy 10^7 to 10^8 years ago.* There is no proof of this hypothesis and the observation of motions in the Galaxy which could be the result of an explosion give an energy of $W_{expl} \sim 10^{56}$ ergs. Secondly the nonstationary model is not supported by the cosmic-ray data. The study of meteorites leads one to conclude that the cosmic-ray intensity has been constant during the last 10^9 years (within a factor of 2; cf. Ref. 3). The existence in cosmic rays of electrons with energy $E \sim (2 \text{ to } 3) \times 10^{10}$ eV shows that they could not have been generated for more than

$$T_{max} \simeq \frac{1.6 \times 10^{13}}{E_{eV}(H^2/8\pi + w_{phot})} \sim 6 \times 10^{14} \text{ sec} \sim 2 \times 10^7 \text{ years}$$

[cf. (2.21) on replacing $H_\perp^2$ by $(2/3) (H^2 + 8\pi w_{ph})$ and using the value $H^2/8\pi + w_{ph} \sim 10^{-12}$ ergs/cm^3]. This is more likely to be an overestimation since in a non-stationary model the average value of $H^2/8\pi + w_{ph}$ is greater than 10^{-12}. However the hypothesis of one great explosion in the Galaxy (1 to 3) $\times 10^7$ years ago is very unlikely.

Let us develop yet another argument. In a nonstationary model almost all the particles are accelerated at the moment of the explosion at a time T_{expl} ago. This implies that they have covered a path vT_{expl} before arriving at the Earth, where v is the speed of the particles. Consequently the nonrelativistic particles for which $v \ll c$ will have covered a path much shorter than the relativistic particles. For this reason, the ratio L/M of the intensity of the L and M groups of nuclei must, in a nonstationary theory, diminish when the energy decreases. In fact as far as we know this ratio increases when the energy decreases (cf. Sec. I.2; in the energy region $\epsilon < 400$ MeV/nucleon the ratio L/M indeed decreases again but this is probably an effect of the ionization losses and other factors). The discussion given here is unfortunately not completely definitive because of the inaccuracy of the measurements or the possibility of supplying an explanation for them based on a nonstationary model. In any case it gives a quite definite impression in so much as no results at all confirm the nonstationary model, which appears to us to be very unlikely.

Let us return to the quasistationary galactic model; let us note that this model is supported by considerations other than energetics. One can show (cf. Ref. 1, Chap. 5) that all the results of cosmic rays near the Earth and in the Galaxy are not in contradiction with this model and can be explained in the framework of it. It is necessary to repeat again, however, that the lack of complete and exact information (chemical composition, spallation probabilities, spectra and composition of the electronic component etc....) and the uncertainty of the structure of the halo and the generation of the nucleo-protonic component in supernovae explosions prevent us from considering the quasistationary galactic model as established without any doubt. It is thus very important to discuss other models. This has already been done with the nonstationary galactic model. We shall pass now to the metagalactic models.

---

*If the explosion had taken place too early $[T_{expl} > (1 \text{ to } 3) \times 10^8$ years ago] the cosmic rays would not have remained in the Galaxy until now.

III.3. ON THE METAGALACTIC MODELS FOR THE ORIGIN OF COSMIC RAYS

In the metagalactic models one assumes that the cosmic rays observed near the Earth are generated outside the Galaxy and penetrate it afterwards. One can discuss several models (cf. Table III). In one of these models one assumes that the energy density of the cosmic rays w_{mg} in the whole metagalactic space is of the same order as the energy density of cosmic rays in the Galaxy, i.e.

$$w_{mg} \simeq w_G \sim 10^{-12} \text{ erg/cm}^3. \tag{3.7}$$

In another model one assumes that the equality (3.7) is only valid for the regions near the Galaxy, like the Local Group or the Local supergalaxy (the existence of such a supergalaxy has not been proved but is possible). Such a local metagalactic model is rather nonstationary and from this point of view is similar to the nonstationary galactic model. The essential difference lies in the fact that in the local metagalactic model one assumes that the source of the cosmic rays is a nearby radio galaxy (or several radiogalaxies), e.g. Centaurus A (distance 3.8 Mpc, i.e. 10^{25} cm).

If the cosmic rays in intergalactic space are isotropic the condition (3.7) is then necessary in the framework of the metagalactic models, at least in the neighborhood of the Galaxy. In fact in the stationary case, Liouville's theorem requires that the cosmic-ray intensity be constant along particle trajectories. It follows from the isotropy conditions of the cosmic rays in the Galaxy and the Metagalaxy* that the energy density of the metagalactic cosmic rays in the Galaxy is $w_{mg,G} = w_{mg}$. It is easy to reach this conclusion also from detailed considerations of the motion of particles passing from the Metagalaxy (field H_{mg}) into Galaxy where the galactic field H_G is much greater than H_{mg}. If one takes into account the nonstationarity, Liouville's theorem is no longer valid and the equality $w_{mg,G} = w_{mg}$ can be affected. But under the conditions in the Galaxy (or in galaxies in general) we cannot imagine any powerful and long-periodic mechanism allowing the cosmic rays to be effectively pumped from the intergalactic space into the Galaxy. It is for this reason that if the conditions (3.7) are not satisfied the only possibility for violating the large value $w_{mg,G} \sim w_G \sim 10^{-12}$ is the existence of a strong anisotropy of the cosmic rays in the intergalactic space. In this spirit one can still discuss a metagalactic model (Table III) in which the cosmic rays have a strong anisotropy in the metagalactic space and an energy density w_{mg} much weaker than the galactic density w_G. In such a model the isotropic cosmic rays in a region where the magnetic field is H_G (the galaxy) pass into the intergalactic space where the field is $H_{mg} \ll H_G$ conserving the adiabatic invariant $\sin^2 \theta/H = \underline{\text{const.}}$ This implies a strong anisotropy in the Metagalaxy $(\theta_{max} \simeq \sqrt{H_{mg}/H_G})$ and one has

$$w_{mg}/w_g = H_{mg}/2H_g.$$

For $H_{mg} \sim 3 \times 10^{-9}$, $H_G \sim 3 \times 10^{-6}$ and $w_G \sim 10^{-12}$ one gets $\theta_{max} \sim 1^0$ and $w_{mg} \sim 10^{-15}$ ergs/cm^3. The question of the isotropy of the cosmic rays in metagalactic space is thus very important and demands a detailed analysis. We will consider this problem in the following paragraph, and we will conclude that anisotropy is impossible. Following this conclusion, the only possibility enabling us to keep the metagalactic model is to satisfy condition (3.7) as we have mentioned. This is exactly what is assumed in the metagalactic models (the latest papers in which these models are assumed are Ref. 22 and Ref. 23). The validity of the relation (3.7) is however very much in doubt and the assumption of the following condition is preferred:

---

*This conclusion is completely analogous to that obtained about cosmic rays far from the Earth when they are observed inside the terrestrial magnetosphere.

$$w_{mg} \ll w_g \sim 10^{-12} \; erg/cm^3. \hspace{3cm} (3.8)$$

Concretely, it is more likely that $w_{mg} \lesssim 10^{-15} \; erg/cm^3$. We shall review briefly (for details, cf. Ref. 1) the arguments for the use of the inequality (3.8).

(1) The kinetic-energy density of this intergalactic gas is $K_{mg} = \rho u^2/2 \lesssim 10^{-14}$ to $10^{-15} \; erg/cm^3$ $[\rho \lesssim 10^{-29} \; g/cm^3, \; u \sim (1 \; to \; 5) \times 10^7 \; cm/sec]$ and the intergalactic magnetic-energy density is $H_{mg}^2/8\pi \lesssim 10^{-15} \; erg/cm^3$ $(H < 10^{-7} \; oersteds)$; one has rather $H_{mg}^2/8\pi \lesssim 10^{-16}$ to $10^{-17} \; erg/cm^3$. It is at the same time unrealistic to suppose that the energy density of the cosmic rays w_{mg} is much greater than $K_{mg} \sim H_{mg}^2/8\pi$. We arrive therefore at the estimate:

$$w_{mg} \lesssim 10^{-15} \; erg/cm^3 \ll w_G.$$

(2) One can estimate the energy density of cosmic rays w_{mg} supplied to the intergalactic space by normal and radiogalaxies. One gets in this way an estimate $w_{mg} \sim 10^{-16}$ to 10^{-17} and we cannot imagine any way of obtaining a value for w_{mg} which would be comparable to w_G. To support this conclusion let us note that there are relatively few strong radiogalaxies and quasars (the nearest quasar 3C273B is 500 Mpc away and one of the strongest radiogalaxy Cygnus A is at 200 Mpc). Let us suppose, for example, that in a sphere of radius 100 Mpc and so volume $V \sim 10^{80} \; cm^3$, there is always a source with the power of Cygnus A supplying 10^{61} ergs of cosmic rays in each explosion. If the explosion is lasting 10^6 years one would have 10^4 explosions during a period of 10^{10} years. Thus 10^{65} ergs are injected into the given volume and even if one doesn't take into account the decrease in energy due to losses by adiabatic expansion of the Metagalaxy, the average energy density of the cosmic rays will be $w_{mg} \sim 10^{-15} \; ergs/cm^3$ (see also Ref. 28).

(3) From radioastronomical data (i.e. the remarkable lack of radio radiation coming from intergalactic space) one can deduce some information about the intergalactic magnetic field H_{mg} and the energy density $w_{e,mg}$ of the electronic component of the metagalactic cosmic rays. Thus, for $H_{mg} \sim 3 \times 10^{-8}$; one has $w_{e,mg} \lesssim 10^{-2} w_{e,G} \lesssim 3 \times 10^{-16}$ ($w_{e,G}$ is the energy density of the electronic component in the Galaxy).

(4) The upper limit established for the flux of cosmic γ rays [cf. (1.17)] allows one to estimate the maximum value of $w_{e,mg}$. Even if one assumes that the energy density of optical radiation in the metagalaxy is $w_{opt} \sim 2 \times 10^{-3} \; eV/cm^3$, one obtains $w_{e,mg} \lesssim (1 \; to \; 3) \times 10^{-2} w_{e,G}$. It is more likely that w_{opt} is of the order of $10^{-2} \; eV/cm^3$ and consequently that $w_{e,mg}$ is smaller than $10^{-2} w_{e,G}$; one gets therefore an estimation for w_{mg} that is identical to that obtained in Sec. (3) above. One gets an even weaker estimation of w_{mg} by considering the metagalactic 3 °K thermal radiation whose energy density is $w_T \simeq 0.4 \; eV/cm^3$. In such a radiation field the energy of the electrons diminishes by a factor of 2 due to the Compton losses in a time T_C of the order of 10^8 years [cf. (2.21) where $H_{\perp}^2$ is replaced by $(16\pi/3)w_T \sim 10^{-11} erg/cm^3$]. The scattering of radio photons of average energy $\epsilon_T = 2.7 \; kT \doteq 7 \times 10^{-4} \; eV$ by relativistic electrons gives x rays (continuous background radiation). Using the data about the isotropic x radiation given in Sec. I.4 one obtains an estimate $w_{e,mg} \lesssim 3 \times 10^{-5} \; eV/cm^3 \lesssim 10^{-3} w_{e,G}$. Here, as previously, one assumes that the electrons and the radiation fill the entire Metagalaxy. However, $w_{e,mg} \lesssim 0.1 w_{e,G}$ even if the electronic

component is contained in a region of dimensions $\sim$ 15 Mpc $\sim$ 5 $\times 10^{25}$ cm $\sim 10^{-2}$ R_{ph} ($R_{ph} \simeq 5 \times 10^{27}$ cm is the photometric radius of the Metagalaxy). And in this way even for a region of dimensions greater than 10 Mpc

$$w_{e,mg} \ll w_{e,G}. \tag{3.9}$$

This implies that in the metagalactic models, except for the extremely local model, it is necessary to assume that the electronic component is formed in our Galaxy. However, in this case the metagalactic model for protons and nuclei origin also becomes less plausible (see later).

(5) From the data on the continuous background of x radiation and the state of intergalactic gas, one can conclude, after making a few supplementary assumptions (cf. Sec. III.4), that $w_{mg} \lesssim 10^{-15}$ erg/cm^3.

(6) The value $w_{mg} \sim 10^{-17}$ to 10^{-15} erg/cm^3 is not small, not only in comparison with K_{mg} and $H_{mg}^2/8\pi$ but also in comparison with the kinetic energy contained in the random motion of all the galaxies.

(7) Not only is there no data again of the inequality (3.8) and in favor of the assumption $w_{mg} \sim w_G$ but there is also no real argument in favour of this conclusion (that $w_{mg} \sim w_G$), except the fact that it is possible in this case that the cosmic rays contained in our Galaxy and in other normal galaxies (but not in radiogalaxies and quasars) come from the Metagalaxy. Apart from other difficulties, such a metagalactic theory cannot probably explain the chemical composition of cosmic rays (the metagalactic cosmic rays produced during the stage of the galaxies formation would probably be poor in heavy elements; the cosmic rays coming from galaxies and quasars are most probably insufficient to give a cosmic-ray energy density of $w_{mg} \sim w_G$).

To summarize, one reaches the conclusion that in the framework of metagalactic models it is necessary practically to limit oneself to the discussion of the nuclear component and to consider the electronic component as coming from the Galaxy. However such an eclectic scheme is not very attractive. In actual fact, according to the development of Sec. III.2 it is clear that the theory of the galactic origin of the electronic component of cosmic rays is related to the hypothesis of the existence of the rather large halo and the quite powerful injection of electrons for example, by supernovae explosions. The only step to make between this hypothesis and the galactic hypothesis for all cosmic rays is to postulate an injection of protons and nuclei 10 to 100 more powerful than the injection of electrons. One knows already from the example of the Sun that the generation of the nuclear component is more efficient than for the electronic component. One can also give other arguments in the same spirit that have already been partly mentioned. Let us note also that the local metagalactic model would be nonstationary. Thus everything opposes the metagalactic models, however they have their supporters (Ref. 22 and 23) and certainly some of them cannot yet be considered as definitely refuted. Much effort is still needed to resolve completely the problem of the relative merit of different theories for the origin of cosmic rays (last discussion see at Ref. 32 and 33).

III.4. COSMIC RAYS AND PLASMA PHENOMENA*

The possibility of understanding numerous problems as yet unsolved in the astrophysics of cosmic rays is, we feel, tied to the study of phenomena occurring in a plasma. Of course, the plasma nature of the cosmic milieu is well recognized and this opinion is thus quite an obvious one–it is mentioned in Ref. 1 and in numerous other texts. Here we shall offer a few more concrete comments, tied to beam

*This paragraph is in fact an Appendix and its conclusions were already used in Secs. III.2 and III.3.

instabilities and other instabilities characteristic of tenuous plasmas, that are applicable to the problems that concern us (Ref. 6 and 10).

  Let us discuss such possibility: the magnetic field of a galaxy passes out into the intergalactic space in a continuous manner (expansion of the tubes of force) without any inhomogeneities, shock-waves, etc. In such conditions the isotropic or quasiisotropic cosmic rays will move in the galaxy with conservation of the adiabatic invariant and form in the Metagalaxy a beam of particles moving along the magnetic field (cf. Sec. III.3). However, in such a case a beam instability will develop. In effect the plasma frequencies in the Metagalaxy and in the beam are much greater that I/T_G where T_G is the characteristic time of evolution of the galaxy or even the characteristic explosion time of the radio galaxies. We shall present without further clarification, some parameters of the metagalactic plasma (cf. for example Ref. 24) of density $n \sim 10^{-5}$ cm^{-3}. In this case the plasma frequency is $\omega_0 = \sqrt{4\pi e^2 n/m} = 5.64 \times 10^4 \sqrt{n} \sim 10^2$ sec^{-1}, and the Debye radius is $D = \sqrt{\kappa T/4\pi e^2 n} \simeq 5 \sqrt{T/n} \sim 5 \times 10^5$ cm (for a temperature of $T \sim 10^5$ °K). In these conditions the number of collisions between ions and electrons is $\nu = (5.5n/T^{3/2})$ $\times \ln (220T^{-1/3}n) \sim 10^{-11}$ sec^{-1}. In a field $H_{mg} < 10^{-7}$ the gyromagnetic frequency is $\omega_H = eH/mc = 1.76 \times 10^7 H < 10$ sec^{-1} and consequently $\omega_0^2 \gg \omega_H^2$.

  One can see from these numbers that plasma waves can propagate with very little attenuation in the intergalactic space for $kD = (2\pi/\lambda) D \ll 1$. Later, even for a cosmic-ray density of $N_{CR} \sim 10^{-13}$ cm^{-3} (in the Galaxy, $N_{CR} \sim 10^{-10}$, the value of $N_{CR} \sim 10^{-13}$ in case of identical spectrum corresponds to $w_{CR} \sim 10^{-15}$ erg/cm^3) the plasma frequency of the beam is taken as $\omega_s = \sqrt{4\pi e^2 N_{CR} c^2 / E} \sim 5 \times 10^4$ $\times \sqrt{N_{CR} m/M} \sim 5 \times 10^{-4}$ (here the total energy is $E \sim Mc^2 \sim 10^9$ eV, which corresponds to those protons which give the dominant contribution to w_{CR}.) It is clear that the ratio $2\pi/\omega_s T_G \sim 10^4/T_G$ is very small. The fact that the ratio $1/\gamma T_G$ (where γ is the amplification factor-increment for the plasma oscillations that develop as a result of the stream instability) is small, is even more important. The estimation of the value of γ for the case under consideration leads to $1/\gamma_{max} \sim 30$ years if one assumes that the random velocities in the beam are of the order of c. In fact one has $\gamma_{max} \sim \omega_s^2/\omega_0$, the largest increment corresponding to $kc \sim \omega_0$ (for the shortest waves, weakly attenuated, one has $kv_T \sim \omega_0$, $v_T^2 \sim \kappa T/m$ or $kD \sim 1$ which leads to $\gamma_{min} \sim (\omega_s^2/\omega_0)(v_T^2/c^2) \sim 10^{-4}\gamma_{max}$ i.e. $1/\gamma_{min} \sim 3 \times 10^{+5}$ years). If one takes the magnetic field into account, the beam will also create waves of a different type, the first effect of which is to increase the rate of amplification of the perturbations. Besides even in the absence of a field, the beam instability does not lead only to the generation of plasma waves. Finally at the edge of the Galaxy (and especially for radiogalaxies) one can hope to have a particle density N_{CR} several orders of magnitude greater than the maximum probable value ($N_{CR} \sim 10^{-13}$) that was used for the Metagalaxy.

  Consequently, one has the impression that cosmic rays originally anisotropic must very rapidly become isotropically distributed as a result of the increase of the amplitude of the oscillations. One cannot, unfortunately, draw definite conclusions here (at least as far as the beam instability is concerned) because of the lack of knowledge about the nonlinear phase of the process and also because of the lack of data permitting us to study the influence on the cosmic beam of waves generated by different sources.

  We shall clarify these remarks.

  One assumes in general (and with some justification in the case of a beam of nonrelativistic particles) that the beam instability leads to the formation of a plateau in the velocity distribution of the particles (in the one-dimensional cases one knows that the beam is unstable when the velocity distribution shows a maximum for $v \neq 0$; this is why the instability disappears with the creation of a plateau). At the time of formation of the plateau, plasma waves are generated with total energy comparable to the initial energy of the beam. Apart from the tendency to create a plateau, the beam tends to become isotropic because the anisotropic velocity distribution is, in fact, unstable whereas for an isotropic distribution even the presence of a maximum in the energy (or velocity) spectrum does not lead to an instability. From this point of view one must expect, due to the mentioned very large values of the factor γ of amplification of the beam instability, a very rapid evolution of the cosmic rays to isotropy and the creation of a plateau in the energy spectrum. One cannot assume however, that these conclusions are inevitable, even if the value of γ obtained in the linear approximation is very large. This is because the non-linear interaction between the waves, especially in relativistic beams can lead to a stabilization of the beam (cf. Ref. 25).

From the physical point of view, this is related to processes taking place in the nonlinear approximation such as transformation of waves generated in the beam into the nonresonant part of the spectrum where the waves do not interact directly with the beam (if the phase velocity $v_\phi = \omega/K$ of the waves is larger than c, the waves cannot be absorbed by the particles of the beam). On the other hand the evolution towards isotropy of the beam occurs not only under the influence of waves created by the beam itself but also under the influence of waves (with $v_\phi < c$) produced by other sources present in the region traversed by the beam. In cosmic conditions there exists a number of sources of various waves (beams liberated by stars, or drift produced in the cosmic plasma under the actions of gravitational and magnetic fields, etc.) but their intensity cannot yet be estimated.*

Thus, in the problem of the evolution towards isotropy of the cosmic rays due to the beam instability in an isotropic plasma and under the influence of the plasma waves, there remains a large undetermined factor. But, it seems to us, this indeterminacy concerning the destiny of the cosmic-ray beams disappears on the whole if one takes into account the existence of an aperiodic anisotropic instability which has not yet been mentioned: an anisotropic distribution of particles in a magnetic field is, in general, unstable even if one does not consider the plasma waves. Due to this instability the lines of force of the field bend and naturally the field becomes turbulent. The criterion for this anisotropic instability is the following [there are also other types of instability that appear when condition (3.10) is not satisfied]:

$$w_{CR,\parallel} - \frac{1}{2} w_{CR,\perp} > \frac{H^2}{8\pi} , \qquad (3.10)$$

where $w_{CR,\parallel} = MN_{CR}\bar{v}_\parallel^2/2$, $w_{CR,\perp} = MN_{CR}\bar{v}_\perp^2/2$, bar is velocity averaging and $v_\parallel$ and $v_\perp$ are the components of the velocities of the cosmic rays parallel and perpendicular to the field (we are using a nonrelativistic expression but this does not leave the possibility of considerable error because for the great majority of the cosmic rays one has $E - Mc^2 \sim Mc^2$).

During the passage of the particles from the Galaxy to the Metagalaxy with conservation of the adiabatic invariant, one has $w_{CR,\parallel} \simeq w_{mg} \gg w_{CR,\perp}$ as already shown, while

$$w_{mg} \simeq \frac{H_{mg}}{2H_G} \quad w_G \sim \frac{H_{mg}H_G}{8\pi} \gg \frac{H_{mg}^2}{8\pi} .$$

Consequently the criterion (3.10) will be very well satisfied (for example, for $H_{mg} \sim 3 \times 10^8$, one has $w_{mg} \sim 3 \times 10^{-15}$ and $H_{mg}^2/8\pi \sim 3 \times 10^{-17}$). The amplification

---

*Apart from this, depending on the spectra of the plasma waves or waves of other types, the cosmic rays may be accelerated or slowed down by absorption or emission of these waves. Let us note that because of scattering of these waves on the cosmic rays, in the conditions of interest, the latter are decelerated. This process is analogous to synchrotron or Compton losses. One has $dE/dt = - bw_p E^2$, where b is the same coefficient as appears in the Compton and synchrotron losses and w_p is the energy density of the plasma waves. Since we always have $H^2/8\pi \gg w_p$ in the cases of interest, the synchrotron losses are more important than the losses by scattering on the plasma waves.

factor for the perturbations of the field is $\gamma_a \sim \sqrt{(w_{mg} - H^2 8\pi)/MN_{CR}} \cdot k \sim ck$, because $w_{mg} \sim N_{CR} Mc^2$. The wavelength of the perturbations must exceed the radius of gyration of the cosmic rays

$$r_H \sim \frac{Mc^2}{eH} \sim \frac{3 \times 10^6}{H_{mg}} \sim 10^{14} \text{ cm} \quad \text{(for } H_{mg} \sim 3 \times 10^{-8}\text{)}.$$

Consequently $\gamma \lesssim 2\pi c/r_H \sim 10^{-3}$ and $1/\gamma_{a,max} \sim 10^3$ sec (!). The plasma waves that appear due to the beam instability can clearly only accelerate the evolution towards isotropy of the beam, and make the field turbulent.

It is clear then that if a beam of cosmic rays would arrive in the metagalactic space (a region where $H \sim H_{mg} \lesssim 3 \times 10^{-8}$) without in the beginning evolution to isotropy and without the destruction of the ordered nature of the field as assumed above, these processes will take place very quickly. This implies that the tubes of force do not expand regularly and that, in fact, there exists a transition zone, between the regular field of the galaxy and the metagalactic field where the field is turbulent and the cosmic rays become isotropic. Consequently, if a large quantity of cosmic rays is generated in a galaxy (for example, due to a galactic nucleus explosion) these cosmic rays will rapidly isolate themselves—that is, the region in which they are contained will be surrounded by a turbulent envelope which will prevent them from passing into the exterior region where the field is weak. *

For this reason, one observes, in radio galaxies, the radio radiation of a clouds and not a flowing of cosmic rays along the lines of force of the field. Naturally these clouds, containing isotropic cosmic rays, can overflow and be displaced en bloc as it is in actual fact observed. In addition the diffusion of the cosmic rays out of the system (galaxy, "clouds" in the radio galaxies, supernovae envelopes) is equally probable even in the presence of a turbulent envelope.

Unfortunately an accurate estimate of the diffusion coefficient D_{CR} in a turbulent envelope is not possible without a detailed model and without taking account of the important instabilities and the nonlinear interactions of the waves. One can take as a minimum for D_{CR} the value $D_{CR,min} \sim cr_H/3 \sim Mc^3/3eH$ (taking $E \sim Mc^2$ for the particle energy). Thus for the Galaxy one obtains the value $D_{CR,min} \sim 10^{22}$ cm^2 sec^{-1} (for $H_G \sim 3 \times 10^{-6}$) which is much smaller than the coefficient $D_{CR} \sim 10^{28}$ to 10^{29} that was used in Ref. I (see also Sec. III.2). There is however no contradiction here because in the turbulent region of the galactic periphery the field is weaker than in the Galaxy in average and the coefficient D_{CR} may be of course greater than $D_{CR,min}$.

In any case, the appearance of a turbulent envelope in the transition region between a strong and a weak field is a very favourable condition for the confinement of the cosmic rays in supernova envelopes, in the Galaxy, in the regions (clouds) emitting radio waves in the radiogalaxies, etc..... .

In particular one can take the characteristic time for electrons to leave the Galaxy to be of the order to $T_e \sim (2$ to $3) \times 10^8$ years, such a value not being in contradiction with the chemical composition of the cosmic rays (cf. Sec. III.2).

As it has been seen, one can expect an evolution towards isotropy[†] of the cosmic rays due to the conjugate action of various instabilities of an anisotropic distribution of the particles; in so much as the progress towards isotropy is not complete, the distribution function have a tendency to make a plateau.[‡]

---

*When one considers the spiral arms of a galaxy and in particular the Galaxy the mentioned processes would give the formation of a turbulent "cork" which prevents the cosmic rays from leaving the arms. Naturally, the turbulent envelope forms gradually and a fraction of the cosmic rays leave the system without difficulty.

†In the nonhomogeneous case the isotropization cannot be complete in the sense that diffusion of course must take place and thus there is some diffusion flux.

‡We are not considering here the solar cosmic rays because the departure from the system is then very fast and the evolution towards isotropy possibly does not have sufficient time to take place. Nevertheless, even for the solar cosmic rays it is necessary to take the above considerations into account.

One has seen, then, that the cosmic rays must have become very rapidly isotropic in the course of their passage through the peripheral zone of the galaxy. For
this reason alone, without considering the possibility of an evolution towards isotropy in the intergalactic space, the metagalactic cosmic rays must be isotropic as
was assumed in Sec. III.3.

The evolution towards isotropy of the cosmic rays created, for example, during
the explosion of galactic nuclei and thus anisotropic during a certain time (this
always occurs during the regular flaring of cosmic rays in a region where there is
a weak ordered field) is accompanied by the generation of various types of waves.
The total energy contained in these waves (in particular in the plasma waves) is
probably in order of magnitude equal to that contained in the cosmic rays. During
the propagation of the waves in the intergalactic space, these waves heat up the
intergalactic gas by collisions. The gas is also heated as a result of ionization
losses by the subcosmic rays (particles of energy $E_k = E - mc^2 < 10^8$ eV) whose
existence is very likely. The calculations in Ref. 10 lead to the conclusion that one
has approximately $\kappa n T \sim w_{mg} + w_{SCR}$ (here T is the temperature of the intergalactic gas of concentration n, and w_{mg} and $w_{SCR} \equiv w_{mg,SCR}$ are, respectively,
the energy density of the cosmic rays and the subcosmic rays in the intergalactic
space). One deduces that for $n \sim 10^{-5}$ and $w_{mg} + w_{SCR} \sim 10^{-15}$ to 10^{-16} erg/cm^3
the temperature is $T \sim (w_{mg} + w_{SCR})/\kappa n \sim 10^5$ to 10^6 °K. Consequently, in the
framework of an evolutionary cosmology (the calculations of Ref. 10 are based on
this assumption) the temperature of the intergalactic milieu can be and must be
raised despite the cooling due to expansion of the Metagalaxy. Another argument
is more important for this course: if the condition (3.7) is satisfied (that is, if the
energy density w_{mg} is of the order of $w_G \sim 10^{-12}$) when the cosmic rays are produced in the radio galaxies, as is generally assumed in the metagalactic theories
of the origin of cosmic rays (cf. Ref. 22), then one can believe that the heating of
the intergalactic gas brings it to the temperature $T \sim w_{mg}/\kappa n \sim 10^9$ °K. From
the data on the continuous background x radiation one can deduce a temperature
$T < 3 \times 10^6$ (if one uses the estimate $T \sim w_{mg}/\kappa T$, one gets $w_{mg} < 3 \times 10^{-15}$).
The measurements of the background x radiation above 50 Å would give a more
accurate estimate of T if T is greater than 4×10^5 °K. The use of the formula
$T \sim w_{mg}/\kappa n$ and the data on the x radiation is related to a few supplementary
assumptions. In any case, one obtains by this way another argument in favour of
the inequality $w_{mg} \ll w_G \sim 10^{-12}$ erg/cm^3.

The study of plasma effects in cosmic rays astrophysics is just starting and involves
considerable difficulties. It is for this reason that the discussion is not sufficiently
developed. It would be better to talk of hypothesis and program of study rather
than of well-defined conclusions. Nevertheless we believe that the discussion of
plasma phenomena has a fundamental value in the future development of the astrophysics of cosmic rays (cf. Refs. 6, 10, 19 and 35).

III.5. FINAL REMARKS

Contemporary astrophysics has in its possession, on one hand, large amounts
of information (concerning celestial mechanics, stellar atmospheres, the structure
of the moon and the planets, etc. . . . .), but on the other hand it suffers from fundamental problems which are not as yet solved. In the first place, there is the problem
of cosmology—the problem connected with the evolution of the Metagalaxy and with
its "global" structure. In fact, it is a question of choosing a model within the framework of relativistic cosmology (such as models with a mathematical or physical
density singularity, oscillating models, models with maximum density, see
Refs. 26 and 27).

The difficulties are due to the fact that there is very little information about
the past, that is to say, the previous state of the Metagalaxy. We will illustrate this
fact using as a parameter the average density ρ of the Universe. At the present,

the most probable value of this density is $\rho = \rho_{mg} \sim 2 \times 10^{-29}$ g/cm^3. This value has not been proved, but if we accept it we can draw the following conclusions. The farthest known objects (i.e. the QSS 3C9 and some others) have a red shift $z = (\lambda - \lambda_0)/\lambda_0 \simeq 2$ (the wavelength of the observed line is $\lambda = (1 + z)\lambda_0$, where λ_0 is the wavelength at which the atom emits). The mean density in this phase of the evolution is $\rho = \rho_{mg}(1 + z)^3 = \rho_{mg}(\lambda/\lambda_0)^3 \sim 5 \times 10^{-28}$ g/cm^3. The thermal radiation at $T = 3$ °K which has been recently discovered has undergone a red shift of about 1000; to this value corresponds a mean density $\rho \sim 10^{-20}$ g/cm^3. The clusters of galaxies were formed at a later date when the density was around 10^{-25} to 10^{-26} g/cm^3. We have not direct evidence, however, regarding the epoch when ρ was $\gg 10^{-20}$ g/cm^3. At the same time for the evolving Universe, one accepts a state with nuclear density $\rho \sim \rho_n \sim 10^{14}$-$10^{15}$ g/cm^3, and even $\rho \gg 10^{15}$ g/cm^3 in the cases when a "physical" singularity is involved.

This obviously represents a large extrapolation and the cosmological problem is far from being solved. We can say the same thing about another fundamental astrophysical problem, that is the explanation of QSS's and the explosions in radiogalaxies and in supernovae. Furthermore, we have seen that the problem of the origin of cosmic rays is very much related to the cosmological problem (better yet to the more general problem of extragalactic astronomy) and to the problem of explosions in supernovae and in the nuclei of galaxies. Consequently, we want to emphasize that the uncertainty regarding the origin of cosmic rays is not a particular case. It is in fact only an aspect of the present state of astronomy and is thus related to the existence of unsolved fundamental problems. We may compare the importance of these problems, and especially of the cosmological problem, to problems in physics concerning the structure of elementary particles and to the nature of their interactions.

One may ask what is the essence of the above remarks? We would first nevertheless like to argue against the scepticism encountered (especially among physicists) concerning the problem of the origin of cosmic rays and other astrophysical problems. Many may not as yet be clear but one has nevertheless to be astonished by the progress achieved in this field. The characteristic cosmological distance $R_{ph} \sim 10^{28}$ cm is 15 orders of magnitude (!) larger than the distance between the Sun and the Earth. By comparison, in elementary-particle physics, we have succeeded after tens of years of hard work to come near dimensions of the order of 10^{-14} and 10^{-15} cm, and this is "only" 6 or 7 orders of magnitude smaller than the atomic dimensions $\sim 10^{-8}$ cm familiar to us for a long time. Obviously such arguments are not very relevant but they allow us to consider that modern astronomy is sometimes underestimated in the same manner that modern physics was reproached for its non rigorous use of mathematics.

The above do not increase our understanding but help us to take a more tolerant position toward hypotheses, at times strange and extraordinary which are encountered in astronomy. We have particularly in mind 3 hypotheses which are important from the point of view of these lectures:

    (1) steady-state cosmology,
    (2) local theory of QSS's,
    (3) the theory of the metagalactic origin of the cosmic rays observed near the Earth.

The steady-state cosmology is based on the assumption of continuous creation of matter at the rate $3H\rho_{mg} \sim 10^{-46}$ g/cm^{3}sec. ($H \simeq 100$ km/sec Mpc $\simeq 3 \times 10^{-18}$ sec^{-1} is the Hubble constant, $\rho_{mg} \sim 2 \times 10^{-29}$ g/cm^3 is the mean density of matter in the Metagalaxy). Such an assumption, although very strange, is not easy to reject; this fact is characteristic of the cosmological problem and of its understanding.

The steady-state model was first presented in 1948 and it is only now that it appears to be in contradiction with observations (i.e. the discovery of evolutionary effects with regard to extragalactic radio sources—in the past they were more numerous than at present—and the discovery of the metagalactic thermal radiation).*

The local theory of QSS's presupposes their ejection with speeds comparable to the velocity of light, from the nucleus of our Galaxy or from the nuclei of nearby radiogalaxies. This theory meets with large difficulties as regards energy considerations. Furthermore, in this model one should observe in general many more blue shifts than red shifts; however, in all the known cases (there are more than 50) one observes red shifts. The absence of blue shifts and some peculiarities in the spectra (absorption lines)† if they are correct, as we hope, would allow us to reject with certainty the local theory of QSS's.

We have already discussed the theory of the metagalactic origin of cosmic rays in Sec. III.4. This theory, like the two preceding ones, appear to us very improbable.

Consequently, the author of this course has (always) opposed the "steady-state theory", the local theory of QSS's, and the metagalactic origin of cosmic rays, but his opposition is not an a priori opposition. The fact that these theories are published in scientific journals reflects the present state of extragalactic astronomy. For this reason, to scrutinize and if possible to reject very improbable model is not a waste of time because it permits a better understanding of the situation.

Fortunately, the development of astronomy—the appearance of numerous new methods and the increased possibilities for observations and experimentations—opens new paths towards the verification of new assumptions and models. Therefore there is less and less reason to believe that "everything is possible" in astronomy because it is very difficult to reject theories, even the more extravagant ones. In particular in the field of cosmic rays, we will perhaps be able in the near future to obtain answers to some essential questions at present being discussed. In other words, we have reason to be optimistic concerning the future of cosmic-ray astrophysics.

REFERENCES

1. V. L. Ginzburg and S. I. Syrovatskij, Origin of cosmic rays (Pergamon Press, Inc., 1964).
2. L. I. Dorman, "Cosmic-ray variations and space research", Acad. Sci. USSR, Moscou (1963).
3. Proc. 9th Intern. Conf. on Cosmic Rays (London, 1965), I (1966).
4. V. L. Ginzburg, "Some elementary processes which are important for cosmic-ray astrophysics, and for gamma- and x-ray astronomy", Lecture notes, Summer School at Les Houches (1966).
5. W. R. Webber, "The spectrum and charge composition of the primary cosmic radiation", Handbuch der Physik 46,2 (1966).
6. V. L. Ginzburg and S. I. Syrovatshij, Proc. Intern. Conf. on Cosmic Rays (London, 1965) I, 53 (1966). Usp. Fiz. Wauk 88, 485 (1966). Engl. transl. Sov. Phys.-Uspekhi.
7. B. M. Kuzhevsky and S. I. Syrovatskij, JETP 49, 1950 (1965), Engl. transl. Sov. Phys.-JETP.
8. R. R. Daniel and N. Durgaprasad, Prog. Theoret. Phys. 35, 36 (1966).
9. K. Greisen, Phys. Rev. Letters 16, 748 (1966).
10. V. L. Ginzburg, Astron. J. USSR 42, 1129 (1965); see also 42, 943 (1965).

---

*The proponents of the steady-state cosmology till now attempted to interpret observations so that they are not in conflict with their models. We can hope that this problem will be finally resolved very rapidly.

†One can in principle, estimate from these lines the distance of the QSS's. If this distance is ≫ 10 Mpc the local theory can be rejected. Let us mention also that we do not see for the QSS at cosmological distances any principal difficulties connected with the variations of QSS radioemission (see Ref. 29).

11. P.S. Freier and C. J. Waddington, J. Geophys. Res. 70, 5753 (1965).
12. E. T. Byram, T. A. Chubb and H. Friedman, Science 152, 66 (1966); H. Gursky
    et al., Astrophys. J. 144, 1249 (1966).
13. V. L. Ginzburg, Usp. Fiz. Nauk 89, 549 (1966) [English transl.; Sov.
    Phys.—Uspekhi].
14. V. L. Ginzburg and S. I. Syrovatskij, Ann. Rev. Astron. and Astrophys. 3, 297
    (1965).
15. B. Y. Mills, Ann. Rev. Astron. and Astrophys. 2, 184 (1964).
16. B. H. Andrew, Monthly Notices Roy. Astron. Soc. 132, 79 (1966).
17. B. Y. Mills and J. R. Glanfield, Nature 208, 10 (1965).
18. S. B. Pikelner, Dokl. Akad. Nauk SSSR 88, 229 (1953).
19. E. N. Parker, Astrophys. J. 142, 584 (1965).
20. J. E. Felten, Astrophys. J. 145, 589 (1966).
21. S. I. Syrovatskij, Astron. J. USSR 43, 340 (1966), JETP 50, 1133 (1966).
22. E. M. Burbidge and G. R. Burbidge, Proc. Intern. Conf. on Cosmic Rays
    (London, 1965), 1, 92 (1966).
23. R. J. Gould and W. Ramsay. Astrophys. J. 144, 587 (1966).
24. V. L. Ginzburg, Propagation of electromagnetic waves in plasma (Pergamon
    Press Inc., 1964).
25. V. N. Tsytovich and A. D. Shapiro, Nucl. Fusion 5, 228 (1965).
26. Ya. B Zeldovich, Adv. in Astron. and Astrophys. 3, 241 (1965).
27. V. L. Ginzburg, Astronautica Acta 12, 136 (1966).
28. V. L. Ginzburg and S. I. Syrovatskij, "Cosmic rays in the Galaxy," Report at
    the I.A.U. Symposium No. 31 (Holland, 1966) Published in Ref. 31, p. 411.
29. V. L. Ginzburg and L. M. Ozernay, Astrophys J. 144, 599 (1966).
30. Proc. 10th Intern. Conf. on Cosmic Rays. Calgary, Canada (1967). Partly
    published in Canadian Y. Phys. 46, N 10, p. 2-4 (1968); invited papers were
    published in separate volume by University of Calgary, Canada.
31. Radio Astronomy and Galactic System. IAU Symposium N31. Ed. by H. van
    Woerden. Academic Press (1967).
32. V. L. Ginzburg and S. I. Syrovatskij. See Ref. 30, pg. 48 and this book p. 326.
33. V. L. Ginzburg, Astrophysics and Space Science, 1, 125 (1968).
34. K. C. Anond, R. R. Daniel and S. A. Stephens. Phys. Rev. Letters 20,
    764 (1968); Proc. Indian Acad. Sci. 57, 267 (1968).
35. V. L. Ginzburg and S. I. Syrovatskij. Astrophysics and Space Science. 1,
    442 (1968).

PART II

INTRODUCTION

The following paper is the seventh in a series which we (or one of us) have pre-- sented before the International Conference on Cosmic Rays, beginning from 1955 (see [1-6]). Twelve years is a rather long period of time and it would therefore be interesting to review the development of ideas about the origin of cosmic rays over this period (see [1-6] as well as [7-11]). In so doing two facts seem to stand out. On the one hand, we see indubitable progress in the study of primary cosmic rays and in branches of astronomy concerned with the problem of the origin of cosmic rays. On the other hand some of the most basic elements forming the basis of the most probable galactic model of cosmic ray origin are still not clear enough or are not proven with sufficient rigor. For this very reason it seems that in each and every paper we keep coming back to the same questions on metagalactic cosmic rays, galactic halos, the sources of cosmic rays within the Galaxy, and so on. This type of situation cannot but bring about a feeling of dissatisfaction, particularly among physicists. In this connection we would like to emphasize that the lack of definiteness of the foundations of the theory of the origin of cosmic rays which we have noted is primarily a reflection of the prevailing state of galactic and extra-galactic astronomy. In these fields of astronomy there is much that is still not clear, and to rigorously prove the validity of a number of ideas would be extremely difficult.† In this respect an outstanding example would be the problem of the nature of quasars. The red shift in the spectral lines of these objects may in principle be explained in one of three ways:

A. As the participation of the quasars in the general expansion of the Metagal-axy (the cosmological hypothesis);

B. As the ejection of quasars with an appropriate velocity from the nucleus of the Galaxy or from the closest radiogalaxies and;

C. As the gravitational shift of the lines emitted by a gas occupying the central part of a cluster of neutron stars [13]. Of these hypotheses, only the cosmological hypothesis appears to us as well as to most astronomers and physicists to be prob-ably true. However, not only has this not yet been successfully proven, but in one of the latest papers on this subject (see [13]) it has been asserted, contrary to what

---

†It would appear that from the point of view of the verifiability of a theory and of the possibility of proving the validity of various different hypotheses, the situa-tion in physics is more favorable in most cases. However, the history of the study of superconductivity and of the verification of the general theory of relativity (see for example [12]), as well as of certain hypotheses in elementary particle phys-ics illustrate with sufficient clarity the difficulties that exist with regard to the verification of theories and hypotheses even in the field of physics.

has been accepted, that quasars are distributed no further than 40 mpc from the Galaxy. The arguments in support of this position do not appear to us to be sufficiently based on observation. One cannot, however, disagree about the uncertainty prevailing with respect to the remoteness of the quasars. If quasars actually had a local character and if it turned out, in fact, that they were indeed ejected from the galactic nucleus (see [14]), then our conceptions about the structure and history of the Galaxy would undergo a radical change. The same would possibly also hold true in such case with respect to the problem concerning the origin of galactic cosmic rays.

Because of this situation continuation of the discussions of the basic characteristics of the various models used for the origin of cosmic rays is inevitable. This by no means implies that the different models are to be considered equally valid. On the contrary, we consider the most probable model to be the galactic model with a halo [1-11]. But until the validity of such a model is proven, the analysis of alternative possibilities remains one of the most important tasks.

Thus, discussions concerning the fundamental questions about the origin of cosmic rays is still essential. In view of the difficulty in the solution of the astronomical problems that are involved here, this necessity should be of no surprise.

However, another question arises, has a sufficient amount of new material been accumulated during the two years since the last conference to make it worth our while to dwell on this subject? In this respect we are not completely sure. However, inasmuch as we now have available new data, evaluations, and ideas, we hope that their presentation will not turn out to be unnecessary and that it will facilitate a fruitful discussion of relevant problems at this conference.

I. METAGALACTIC COSMIC RAYS

(The Uniform Model)

In what follows, when we speak of the origin of cosmic rays we shall refer to the main part of cosmic rays, observed on the Earth. Corresponding to these particles (with an energy, say, of $E < 10^{15}$ - 10^{17} electron-volts) the energy density near the solar system is of the order $w_G \sim 10^{-12}$ ergs/cm³.* In Galactic theories (or models) the value of w_G is determined by particles formed inside our Galaxy; according to the metagalactic theories, however, the sources of the cosmic rays of interest to us are distributed outside of the Galaxy. In dealing with metagalactic models the energy density for metagalactic cosmic rays† would be equal to

$$w_{Mg} \sim w_G \sim 10^{-12} \text{ erg/cm}^3 \qquad (1)$$

*Integrating over the spectrum of the cosmic rays observed from the earth in a period of minimum solar activity yields the value $w_G = 0.6$ eV/ cm³.

†The nonobservance of the relation (1) in metagalactic models would take place for anisotropic metagalactic cosmic rays. This case, however, appears improbable (see [6, 15, 16]). Another possible factor that could give rise to the inequality $w_{Mg} \ll w_G$ is tied in with considerations involving the nonstationary character of the problem. In this regard, one may cite, in particular, the model brought forth in [17], in which the cosmic rays within the Galaxy would be compressed by gas clouds from intergalactic space falling onto the Galaxy (This type of cloud collapse could continue on only for a quite limited amount of time). We do not see, however, any possibility of obtaining in this manner any extreme violation of (1) for any sufficiently prolonged period of time. In fact, even in [17], the question brought up was only the possibility of decreasing the value w_{Mg} by one order of magnitude.

If the evaluation in (1) holds true for all metagalactic space (or more exactly for the region whose dimensions R are of the order of magnitude of a photometric radius, where $R_{ph} \simeq 5.10^{27}$ cm) then the corresponding model shall be called a uniform metagalactic model. If, however, the evaluation in (1) refers only to the region in the vicinity of the Galaxy whose dimensions $R \ll R_{ph}$, then we shall refer to it as a local metagalactic model. In the case of a Local Group of galaxies

$$R \sim 10^{24} \text{ cm}$$

In the "Centaurus A model" (see below) $R \sim 10^{25}$ cm; for a hypothetical Local Supergalaxy $R \sim 10^{26}$ cm.

From the foregoing discussion three basic types of models emerge as possible choices. They may be delineated by the following scheme:

| Galactic models |

| Metagalactic models |

| The uniform model | | The local model |

The problem of choosing among them is one of the most fundamental questions in the theory of the origin of cosmic rays. Arguments against metagalactic models were brought forth, in particular, in the preceding paper [6] and we shall not repeat them here. We shall only dwell upon those assertions which may be made more exact.

Based upon facts known about isotropic cosmic X-rays (for a summary of the results see [18]) one may make rather definite conclusions concerning the electron component of the metagalactic cosmic rays. Using facts known on X-rays and assuming the existence of metagalactic thermal radiation whose temperature is equal to $3°K$[†] we may obtain the same type of evaluation for the upper limit of the energy density of the relativistic electrons in metagalactic space:

$$w_{e, Mg} \lesssim 3.10^{-17} \text{ erg/cm}^3 \ll w_{e, G} \sim 10^{-14} \frac{\text{erg}}{\text{cm}^3} \qquad (2)$$

Here, $w_{e,G}$ is the energy density of the electron component of the cosmic rays in the Galaxy. We arrive at (2) by assuming that the intensity of the X-ray background within the interval $1.5 < E_X < 6 keV$ is $I_X = 10$ photons/cm^2 ster. sec., and that the energy density of the metagalactic thermal radiation is $w_T = 0.4$ eV/cm^3 ($T = 3°$ K); and we have also assumed that the radiation accumulates on a path $L = R_{ph} = 5.10^{27}$ cm. Furthermore, we took into account the electrons with $E \geq 7.10^8$ eV responsible for X-rays with energy of $E_X \geq 1.5$ KeV. The value of $w_{e,Mg}$ depends of course on the spectrum of the electrons. But this dependence is quite weak and has practically no influence on the order of magnitude of the evaluation (in [6] it was assumed that $T = 3.5°$, $w_T = 0.7$ eV/cm^3 and the value $w_{e, Mg} < 10^{-5}$ ev/cm^3 was carried out; according to [18], $w_{e, Mg} \sim 10^{-3} w_{e, G}$; both of these results do not contradict the estimate in (2)). In (2) only an inequality sign is present, since it has not yet been proven that the X-ray background is a direct result of the scattering of the relativistic electrons by thermal photons (a contribution to the intensity of the X-ray background could also be made by the

[†]The experimental data appearing in [19] is evidentally consistent with the value $T = 2°K$. This is without mentioning however, that they refer only to the wave region $\lambda \geq 1.5$ cm. In the case where $T = 2°K$ the losses due Compton effect decay are $(3/2)^4 \simeq 5$ times less than the case where $T = 3°K$; thus an accurate determination of the temperature T is quite essential. We shall use the more probable value of $T = 3°K$ (the best value for the begining of 1969 is $T = 2.7°K$).

X-ray radiation of the galaxies and by the Bremsstrahlung radiation of the inter-galactic gas). In addition to the above, a substantial contribution from the region $R > R_{ph}$ may also occur.

Hence the energy density of the electron component of cosmic rays in the Meta-galaxy is at least 300 and probably even 10^3 times less than in the Galaxy near the Earth. This conclusion does not in any way, contradict the information obtained through radio and gamma ray astronomy [6], nor does it contradict the estimates of the amount of relativistic electrons that originate from the Galaxy and find their way into intergalactic space (using the estimate (3) introduced below and taking into account the energy losses, we obtain $w_{e, Mg} \leq 3 \cdot 10^{-18}$ erg/cm^3; the crude-ness of the estimate does not yet allow us to discuss the disagreement with (2) or, more exactly with the estimate $w_{e, Mg} \sim 1 \div 3 \cdot 10^{-17}$ erg/cm^3, which would follow from the X-ray data; see, however, [44]).

The metagalactic model is thus immediately seen to be automatically unfavor-able with respect to the electron component under what amounts to a solitary as-sumption concerning the presence of residual thermal emission at T = 2 - 3°K)* It is clear from here that the preservation of the uniform model for the proton-nuclear component is at the same time tied in with an assumption which postulates a completely different character for the source of this compo-nent and for the source of the electron components. Thus in [20] for example, this particular model is mentioned: The protons and the nuclei are assumed to be of metagalactic origin, whereas the electron component is said to be formed in the Galaxy itself. These types of "mixed" models seem quite improbable to us due to general considerations. In models where the electron and proton-nuclear com-ponents occupy one and the same volume, the total energy contained in the electron component is only 30 ÷ 100 times less than the energy retained by the proton-nu-clear component. However, the energy losses for electrons, generally speaking, is substantially greater than for protons and nuclei. As a result, if we take for example the model for the source of all cosmic rays (see below), it would only be necessary to inject into the proton-nuclear component an energy of about 20-30 times greater than that which would be necessary in order to generate the electron component. (Here it is essential that the electron component in the Galaxy should not be composed largely of secondary particles formed as a result of the decay of $\pi^{\pm}$ mesons). In another respect, in order to explain the known proper-ties of both components the same assumption is necessary. In particular, while it would follow, from the chemical makeup of the nuclear component, that the cos-mic rays pass through a gas thickness of the order of 3 g/cm^2, this value is also acceptable from the point of view of available information concerning secondary (electron-positron) constituents of the electron components of cosmic rays (see for example [21]). We must also add that the facts known about solar cosmic rays as well as a number of theoretical considerations lead to the conclusion that the proton-nuclear component occupies a preferable position in comparison with the electron component during cosmic ray generation at cosmic sources. If even in the very process of generation, protons, nuclei and electrons would all occupy positions of equal importance which occurs for the acceleration mechanism ex-amined in [22], then taking energy losses into account one finds that the cosmic rays emitted by the sources would be impoverished by the electrons (see also Ref. 47)

Leaving this type of argumentation aside we shall mention the fact that there are also many other objections (see in particular [6] [23]). to the uniform model as well as to assumption (1) tied in with this model governing the entire Metagalaxy (for $R \lesssim R_{ph}$). We shall dwell here on only one aspect of the matter: to "fill" the

* In the uniform metagalactic model the electrons must observe the relation $w_{e, Mg} \sim w_{e, G} \sim 10^{-14}$ erg/cm^3.

Metagalaxy with cosmic rays with the density in (L) would be extremely difficult. If one were to evaluate the amount of cosmic rays in galaxies and in radiogalaxies by the usual method, assuming that the magnetic energy $W_m \sim (H^2/8\pi) V$ is the same order as the energy of cosmic rays $W_{cr} \sim 10^2 W_e$ (here W_e is the energy contained by the electron component) then even without taking energy losses and the expansion of the Metagalaxy into account, we would arrive at the estimate

$$w_{Mg} \lesssim 10^{-15} \div 10^{-16} \ \text{erg/cm}^3 \tag{3}$$

In other words, if $W_{cr} \sim W_m$ then the galaxies could not assure the type of injection of cosmic rays into the intergalactic space that would allow relation (1) to be observed. For this reason in order to establish the uniform metagalactic model on firmer ground the assumption is made in [24] that in radiogalaxies

$$W_{cr} \gg W_m \sim \frac{H^2}{8\pi} V \tag{4}$$

Under the conditions in (4) the energy $W_{cr} + W_m \simeq W_{cr}$ is not the minimum possible energy and it may therefore be chosen so as to assure a large value of W_{Mg}. We have had to emphasize on several occasions (see for example [5, 6, 11]) that the values of W_{cr} obtained in this manner for radiogalaxies turn out to be too large. This conclusion becomes especially clear if one were to calculate (see [25]) the energy distribution over a single radiogalaxy that would be needed to assure the observance of relation (L). Radiogalaxies almost without exception all are bright elliptical galaxies, whose concentration in our era has an estimated value of $N_{Eb} \simeq 1.3 \cdot 10^{-4} \ (\text{Mpc})^{-3} \simeq 4 \cdot 10^{-78} \ \text{cm}^{-3}$ (see [26]). Let us assume that every such galaxy has passed through a radiogalactic phase and that the cosmic rays formed in such a galaxy did not experience any energy losses and did not slow down as a result of the expansion of the Metagalaxy. Even under such assumptions, it would be necessary in order for cosmic rays of the density in (L) to be obtained that every bright ellipitical galaxy inject cosmic rays with energy

$$W_{cr} \sim \frac{w_{Mg}}{N_{E,b}} \sim \frac{10^{-12}}{4 \cdot 10^{-78}} \simeq 2 \cdot 10^{65} \sim 10^{11} \ M_{\odot} c^2 \tag{5}$$

Due to the above discussion this value is reduced, by an order of magnitude. Hence, practically speaking, the evaluation in (5) would be preserved even if one were to suppose that all elliptical galaxies, and not just the bright elliptical galaxies, are exploding (the concentration of all elliptical galaxies is $N_E \simeq 10^{-3} \ (\text{Mpc})^{-3}$) [see note]. But the mass of the giant galaxies usually does not exceed the value $10^{12} \ M_{\odot}$ and thus the conversion of energy, of the type described in eq. (5). into cosmic rays would appear to be excluded. The real maximal value of W_{cr} would be in our opinion $W_{cr, max} \sim 10^{61} \div 10^{62}$ ergs. In the case where $W_{cr} \sim 3 \cdot 10^{61} \sim 10^7 \ M_{\odot} c^2$ we obtain the following evaluation of the density W_{Mg}:

$$w_{Mg} \leq W_{cr} \cdot N_{Eb} \sim 10^{-16} \ \text{erg/cm}^3, \tag{5a}$$

Here the inequality sign results from the necessity of taking into account the energy losses and the expansion of the Metagalaxy. Note that the contribution of the quasars to W_{Mg} in view of their small quantity may be completely ignored (here we assume, of course, that the quasars are situated at cosmological distances).

It would be appropriate to mention that the evaluation in (5a) is in accordance with (3). This is what is to be expected when we refrain from using inequality (4). There are a number of considerations that support such an abstention. First all of as is well known the total energy $W_{cr} + W_m$ is a minimum provided that

$$W_{cr} \sim W_m \ (\text{or} \ w_{cr} \sim \frac{H^2}{8\pi}) \tag{6}$$

In the second place in the case of radiogalaxies (excluding the initial stages) as well as in several other cases, condition (6) naturally follows from dynamical considerations. Indeed, if an injection of cosmic rays into some region in a field H takes place, then the cosmic rays will be retained by this region only so long as $w_{cr} \lesssim H^2/8\pi$; if, however, $w_{cr} \gg H^2/8\pi$ then the cosmic rays will more or less freely "flow out" of this system. Thus in the case delineated in (4) one must assume that the radio emitting clouds are scattering freely. One would then expect these clouds to be structureless, i.e. one would expect a quasi-uniform distribution of the cosmic ray concentration and one would expect that there would be no sharp changes in the magnetic energy density $H^2/8\pi$. As a result of this, under conditions (4), the radio emitting clouds could not, it would seem, manifest any fine structure in the intensity distribution of radio emission. However, observations under conditions of high angular resolution tend to support the converse (see for example [27, 28]). In the radio-disc belonging to the Galaxy this tendency—a sharp nonuniformity ("patchiness") in the distribution of radio-brightness-is observed, as is well known, with complete clarity. On the other hand, we know that in the disc the equapartition distribution expressed in (6) is in fact observed.

Thus, it appears that from every point of view galaxies, radiogalaxies and quasars cannot in any way assure the observance of relation (1) in our era (for $R \lesssim R_{ph}$). Within the frame work of evolutionary cosmology only one possibility remains—a powerful injection of cosmic rays at the formation stage of galaxies and quasars (for the sake of definiteness we may assume this takes place when $z = \frac{\lambda - \lambda_0}{\lambda_0} \sim 3 \div 10$, i.e. for $t \sim 3 \div 10.10^8$ years from the conditional inception of the expansion of the Metagalaxy).

We now evaluate the energy density of residual cosmic rays formed at the formation stage of the galaxies, or more precisely, at the stage at which the stars were formed.

The gravitational energy of the system during the formation of the stars decreases; if we set this energy at the pre-star stage equal to zero, then after the formation of a star with mass M and with radius r, the gravitational energy attains an order of magnitude equal to that of $\frac{\kappa M^2}{r}$. Furthermore, it is completely clear that during the process of star formation only the energy $\xi \frac{\kappa M^2}{r}$ may be converted into cosmic rays where $\xi < 1$ and in all liklihood even $\xi \ll 1$. For most stars the energy $\kappa M^2/r \ll Mc^2$: For the Sun, for example, $\frac{\kappa M^2}{r} \sim 10^{-5} \ Mc^2$. A coefficient of 10^{-5} or 10^{-4} may be considered typical for all stars. Furthermore, in the Metagalaxy the mean density of matter concentrated at the stars is $\rho \sim 5.10^{-31}$ g/cm³ or $\rho c^2 \sim 4.10^{-10}$ erg/cm³. From this it is clear that the density of the gravitational energy discharged during star formation is $\sim (10^{-4} \div 10^{-5}) \times .410^{-10} \sim 0.3 \div 3. 10^{-14}$ erg/cm³ and that the part that can be converted into cosmic rays is the energy whose density in our epoch is

$$w_{Mg, r} \sim 0.3 \div 3 \cdot 10^{-14} \xi \frac{erg}{cm^3} \ll 10^{-14} \frac{erg}{cm^3} \tag{7}$$

Even this last estimate depends only upon the quite natural assumption that the inequality $\xi \ll 1$ is valid. In fact, the inequality $\xi \ll 1$ would signify in the formation of a star the energy that would be converted to cosmic rays would be much less than the energy $\kappa M^2/r$. For the sun, $\kappa M^2/r \sim 10^{48}$ ergs and if this amount of energy would be discharged, for example, during a time interval of 3.10^7 years, then this would correspond to a power of $\sim 10^{33}$ erg/sec, which coincides in order of magni--

tude with the complete luminosity of the sun. At the same time the power of the sun as a source of cosmic rays is now $U_\odot \sim 10^{24}$ erg/sec and there is no basis for supposing that this power increases by many orders of magnitude if the protosun is compressed slowly. If, however, we are concerned with the case of rapid formation of protostars, which is what we were in fact concerned with above, then taking the quasi spherical symmetry of the problem into account it would be extremely difficult to conceive of the possibility of realizing conditions for which $\xi > 10^{-2} \div 10^{-3}$ (see also what follows below). Thus it is most likely that

$$w_{Mg, r} \lesssim 10^{-16} \text{ erg/cm}^3 . \tag{8}$$

In what appears above the decrease in energy of the relict cosmic rays resulting from the expansion of the Metagalaxy was not yet taken into account. When this effect is taken into account it leads to a decrease in the density $w_{Mg, r}$ attributed to our era by approximately a single order of magnitude. In this connection estimates (7)-(8) only become even more convincing. We remark finally that the chemical composition of the residual cosmic rays would, in all probability, differ from the observed chemical composition of the rays at the earth. For this reason even if we ignored the energy estimates, which are clearly unacceptable the consideration of residual cosmic rays as a fundamental component of cosmic rays in the uniform metagalactic model would necessarily depend upon additional assumptions (see also [45]).

In summing up we see that the uniform metagalactic model for the origin of cosmic rays meets up with the most serious objections and within the framework of familiar concepts and of evolutionary cosmology this type of model would be unacceptable. It seems to us that one would be more hesitant about reaching such a conclusion only if there would occur a radical change in views in the field of extragalactic astronomy and one were to accept for example the validity of the stationary cosmological model (see also [45]). However, all the contemporary tendencies in the development of astronomy and cosmology seem to be pointed in the opposite direction, (in particular the stationary cosmological model seems more and more improbable if not totally untenable).

II METAGALACTIC COSMIC RAYS

(Local Models)

In local metagalactic models the region filled by cosmic rays with high intensity (condition (L)) has dimensions $R \ll R_{ph} \cong 5.10^{27}$ cm. However, there is no evidence that would indicate any special activity among the galaxies that are distributed within the vicinity of our Galaxy. On the contrary the density of these galaxies and radiogalaxies is in general no greater than the average density. For this reason the energy considerations presented above also apply to many local theories as well. Suppose, for example, that we are dealing with a hypothetical

If we take into account the possible formation of cosmic rays during explosions of galactic nuclei, even then the evaluation in (7) is essentially preserved. Suppose that a fraction of order $\zeta \sim 10^{-2} \div 10^{-3}$ from the entire mass of the galaxies takes part in the explosions of the nuclei. Then neglecting the effects of energy losses and of the enlargement of the region occupied by cosmic rays, $w_{Mg, r} \sim \zeta \xi \cdot \rho c^2 \sim 4.10^{-15} \div 4.10^{-16}$ for $\xi \sim 10^{-3}$. The role played by the enlargement of the region and by energy losses is in this case significant, and for our era it is likely that the evaluation in (8) is again valid.

Local Supergalaxy whose volume is $V \sim 10^{77}$ cm³. In this region we have about 10^4 galaxies. In order for an accumulation of cosmic rays having a density of $w_{Mg} \sim 10^{-12}$ erg/cm³ to occur and even if we ignore the loss of particles by the system as well as the system's expansion each of these galaxies would have to inject cosmic rays with an energy equal to $W_{cr} \sim 10^{61}$ ergs. But this means that every galaxy would have to have passed through a radiogalactic phase and furthermore, would even have to have once been a radiogalaxy of the most powerful type. At the same time as was already em--phasized radiogalaxies are basically if not exclusively bright elliptical galaxies, of which there are very few in the Local Supergalaxy. In quasi stationary local models another difficulty arises related to the trapping of cosmic rays. For this type of trapping the magnetic field must be sufficiently strong, and quasi-closed. However, such an assumption does not have any basis at all, and if we are also interested in explaining the origin of the electron component in the same region of the Local Supergalaxy the assumption would contradict known radio astronomical data (see for example [5]).

In addition to this, when the existence of residual thermal emission with $T = 3°K$ is taken into account, sharp delimitations, independent of the assumptions concerning the nature of the Metagalactic magnetic fields are imposed upon the application of the Local Metagalactic model to the electron component of the cosmic rays. Indeed, during motion over a period of time τ within a radiation field with energy density w_T and with a chaotic magnetic field of intensity H, the energy of an electron is

$$ E \leq E_{max}(\tau) = \frac{1}{\beta\tau} = \frac{1.56 \cdot 10^{13}}{(w_T + \frac{H^2}{8\pi})} \text{ ev}, \quad \beta = \frac{32\pi e^4}{9 \, m^4 c^7}\left(w_T + \frac{H^2}{8\pi}\right), \tag{9} $$

where $w_T + H^2/8\pi$ is measured in ergs/cm³. From this it is clear that in observing an electron in the vicinity of the Earth with energy E it may be asserted that even if the electron moves in rectilinear fashion with a velocity of $v \simeq c$ it may still not transverse a path R larger then

$$ R_{max} = c\,\tau = \frac{4.7 \cdot 10^{23}}{(w_T + \frac{H^2}{8\pi})\,E(ev)} \quad \text{cm} \tag{10} $$

When $w_T = 0.4$ ev/cm³ $= 6.4 \cdot 10^{-13}$ (for $T = 2.7\,°K\ w_T = 4.15. \, 10^{-13}$) and when the magnetobrems losses in the field H are ignored

$$ R_{max} \simeq \frac{7.10^{35}}{E(ev)} \quad \text{cm} \tag{11} $$

Within the makeup of the electron component of cosmic rays in the vicinity of the Earth one definitely observes particles with $E \simeq 3.10^{10}$ ev (the most recent summaries of the available data may be found in [21, 29]) and possibly with energies up to $2 \div 4.10^{11}$ ev (see [30]). In the case of $E = 3.10^{10}$ ev we would have, according to (11), $R_{max} \simeq 2.10^{25}$ cm.

Actually, however, the estimate in (11) is obviously too high. In the first place, it is difficult to conceive of the motion of particles in metagalactic space as rectilinear. Even in a field of $H \sim 10^{-9}$ the radius of curvature of the trajectory of a particle with $E \sim 3.10^{10}$ ev would be $r = E/300H \sim 10^{17}$ cm and therefore even such a field can radically change the trajectory of particles. One would expect that the "by-pass factor" connected with the effect of the intergalactic magnetic field might lower the estimate in (11) by at least a single order of magnitude. In the second place, in traveling from intergalactic space toward the Earth an electron must traverse some path within the Galaxy. Here the energy losses within a unit time interval are about three times greater than in intergalactic space $(w_{tot} = w_T + w_{opt} + H^2/8\pi \sim 2.10^{-12} \text{ erg/cm}^3)$ and the field is more complex. Due to the latter reason the electron will move within the Galaxy (from its "boundaries" to the Earth) for a period of $T > 10^7$ years. When $\tau = 10^7$ years $= 3.10^{14}$ seconds and $w_{tot} = 2.10^{-12}$ erg/sec, then according to (9), $E_{max} = 2.10^{10}$ ev, i.e. the extragalactic electrons with $E > 2.10^{10}$ ev cannot in general reach the Earth. Similar considerations lead one to assume that for sources of electrons with $E \gtrsim 10^{10}$ ev

$$R_{max} \leq \frac{10^{35}}{E\,(ev)} \lesssim 10^{25} \text{ cm.} \tag{12}$$

The distance to the radiogalaxy that is nearest to us, Centaurus A is equal to $R_{CG} = 3.8$ Mps $\simeq 10^{25}$ cm. Thus the sources of the electron component of cosmic rays in the Galaxy would have to have an even more local character in the metagalactic scale (R $\lesssim 10^{25}$ cm $\sim 10^{-3} R_{ph}$; this estimate does not contradict the X-ray data either).

However, this possibility is quite improbable.

For the sake of definiteness let us dwell upon the local metagalactic "Centaurus A model" in which cosmic rays originate from Centaurus A. The path from this source to our position would be traversed within a period $\tau \geq R_{CG}/v_{\parallel} \sim 10^{15}$ seconds ($v_{\parallel} \lesssim 10^{10}$ cm/sec). In the vicinity of the Earth $E_{max} \lesssim 10^{10}$ ev. However, no such sharp cut off in the electron spectrum has been observed. To fill the volume $V \sim R^3_{CG} \sim 10^{75}$ cm^3 with a density $w_e \sim 10^{-14}$ erg/cm^3 Centaurus A would have to inject into the electron component alone an energy of $W_e \sim w_e V \sim 10^{61}$ ergs. In taking energy losses into account the energy that it would be required to inject would be even greater. Finally during the influx of electrons into the Galaxy from without the electrons with high energy would have to radiate energy within, basically, a halo, or on the borders of a radio disc. So the energy density $w_{e,G}$ on the periphery of the Galaxy would have to be greater than on the Earth. Although this problem is of a quantitative character there is no data of a radio-astronomical nature that would support the validity of this sort of picture. On the contrary all of the facts that are known to us are in agreement with the assumption that the energy density and hardness of the electron component at least do not increase with the distance from the center (disc) to the periphery (halo).

According to what was indicated it was considered at the very least unnatural to suppose that the sources of electrons and nuclei (including protons) are not the same and that these sources occupy different regions. If, nevertheless, we apply a local model, specifically the "Centaurus A model", only for the proton-nuclear component then for the sake of the validity of this model one would have to make assumptions that would still be quite improbable. Thus, in order that the volume $V \sim 10^{75}$ cm^3 be filled by an energy density of $w_{Mg} \sim 10^{-12}$ erg/cm^3, Centaurus A would have to inject into the cosmic rays an energy of $W_{cr} \sim 10^{63}$ ergs. At the same time according to customary evaluations (see for example [11]) $W_{cr} \sim 10^{59}$ ergs for Centaurus A. Furthermore, even in the case of gravitational collapse, the energy release does not exceed $0.1 \div 1\%$; for a transformation into cosmic rays one could hardly attain an effectiveness of greater than $10^{-3} - 10^{-4}$. This means that during the collapse of a mass M into cosmic rays an energy conversion of $W_{cr} \lesssim 10^{-3} Mc^2 \sim 10^{51}$ (M/M$_\odot$) ergs takes place. Thus when $W_{cr} \sim 10^{63}$ the mass M $\gtrsim 10^{12}$M$_\odot$ whereas the mass of the entire Galaxy comprising Centaurus A is M $\cong 2.10^{11}$ M $_\odot$

We also note that "the Centaurus A model" would have to be essentially nonstationary. This conclusion has met with many objections (see section 3.1).

In general one may say that the local metagalactic models encounter some of the most serious difficulties. It must be admitted however that they evidently cannot be cast aside with the same degree of definiteness as was the uniform model. We assume, nevertheless, that Local models could attract serious attention only if for instance the hypothesis concerning the local nature of quasars was valid.

More advanced analysis of metagalactic models within the framework of an experimental program ought to be carried out in various directions. Using the methods of gamma ray, X-ray and radioastronomy, available information on intergalactic cosmic rays and particularly the electron component of intergalactic cosmic rays may be made more precise. In so doing, it is to be emphasized that the γ-rays emitted from the disintegration of π° mesons contain information about the protonnuclear component generating the mesons. It follows from a known limit for the flow of observed γ rays (see [31]) that the intensity of cosmic rays within metagalactic space is no greater than within the Galaxy (see [11, 23, 45] for greater detail). For this reason an increase in sensitivity by a single order of magnitude and particularly by two orders of magnitude could directly confirm the validity of the inequality $w_{Mg} \ll w_G$. Unfortunately, to accomplish this increase one would have to know the concentration of intergalactic gas (in what was outlined above we started out from the assumption that in our epoch, the mean metagalactic gas concentration is $n \sim 10^{-5}$ cm^{-3}; however, the question concerning the actual value of n is still open). From what was mentioned above one may clearly see the possibilities which give the further study of the electron component, particularly for energies such as $E > 10^{10}$ ev. The importance of the energy spectrum and of the chemical composition of the cosmic rays for $E > 10^{15} \div 10^{17}$ ev is also clear. In this region, a substantial contribution from the metagalactic component is not only possible but also quite probable. If one could accurately isolate the metagalactic component of cosmic rays for super-high energies, then as a result of well known extrapolations we would also obtain information about metagalactic cosmic rays with less energy. Finally we shall mention the measurement of the degree of anisotropy of cosmic rays

$$\delta = \frac{I_{max} - I_{min}}{I_{max} + I_{min}}$$

Based upon theoretical considerations (see [11]) one would expect in the galactic model an anisotropy of order $\delta \sim 10^{-3}$ whereby the intensity is maximum in the direction of the galactic center. According to available data (for the latest results in this area see [32]) this is precisely the anisotropy that is observed. The effect, however, is so small that the problem can still not yet be considered solved and hence the need for further measurements over a wide range of energies appears quite significant. Conceptually speaking, the measurement of anisotropy is one of the most direct means by which one could distinguish between metagalactic models and galactic models. Indeed in the former case the cosmic rays would have to flow into the Galaxy from outside; the intensity in the direction of the galactic center or in the neighboring direction would have to be minimal. In galactic models, however, the anisotropy is of opposite sign (see above). Unfortunately, a real situation can get significantly more complicated due to the influence of the magnetic field in the galactic region near the solar system. However, it appears to us to be quite improbable that the local field could change even the sign of anisotropy. Thus a reliable determination of this sign, while it would not yield a final solution to the problem, it would greatly facilitate doing so.*

*Whereas different groups of data, particularly those dealing with extensive atmospheric showers, are strongly scattered with respect to the value of δ itself, the direction of maximal intensity in most cases corresponds to the direction of the galactic center or to a neighboring direction. It is possible that this is no accident and that the direction of the maximum is more statistically stable and a more easily determined parameter than the quantity δ.

III. GALACTIC MODELS

III. 1. THE MODELS UNDER DISCUSSION

In galactic models, by assumption, the cosmic rays observed in the vicinity of the Earth (more exactly their principle part) are assumed to be formed in the Galaxy. Distinguishing between the various galactic models reduces primarily to selecting the dimensions of the region filled by cosmic rays and to selecting the sources. The table below outlines the situation.

Model	Region filled by cosmic rays (where $w_G \sim 10^{-12}$ erg/cm^3)	Time dependence	Main sources
Halo Model	Halo ($R \sim 3 \div 5.10^{22}$cm)	Quasi-stationary picture	supernovae, "small" explosions in the galactic nucleus
Disc Model	Disc ($R \sim 5.10^{22}$cm $h \lesssim 2.10^{21}$cm)		
Non-stationary model	?	Great changes during the period $T \lesssim 10^8$ years	"large" explosions in the galactic nucleus

The arguments opposing the hypothesis concerning "large" explosions in the galactic nucleus, and hence against the nonstationary model, have been brought forth more than once [6, 11, 23]. We shall therefore confine ourselves to the remark that a "large" explosion would probably have destroyed the spiral structure of the Galaxy and would have brought it into the category of radiogalaxies. However, only elliptic galaxies are radiogalaxies. The evidence against the "large" explosion hypothesis includes the lack of corresponding variation in the intensity of the cosmic rays, the existence of electrons with high energy (such electrons would not "survive" long enough) and the lack of anisotropic cosmic rays. Of course, if quasars turned out to be ejected from the nucleus of the Galaxy [14] then all these arguments would not be convincing. But this type of hypothesis on the nature of quasars does not seem probable even within the conceptional framework of the local nature of quasars [13].

The halo model and the disc model are distinguished primarily by the choice of volume occupied by cosmic rays. Whereas in the former case this volume would be $V_h \sim R^3 \sim 10^{68}$cm^3, in the latter case it would be $V_d \sim R^2h \sim 10^{67}$cm^3. Correspondingly, the total energy of the cosmic rays in the Galaxy in both models are also distinguished by an order of magnitude:

$$W_h \sim w_G V_h \sim 10^{56} \text{ erg}, \quad W_d \sim w_G V_d \sim 10^{55} \text{ erg} \tag{13}$$

The cosmic ray injection power that would be necessary in order that the quasi-stationary picture be maintained would most probably be higher in the disc model, however, than in the halo model. This is due to the fact that the particle exit time is less in the disc model than in the halo model. Thus, if we make use of a diffusion scheme with an effective diffusion coefficient D then the exit time for the disc model is $T_d \sim h^2/2D \sim 10^5$ years, where the half thickness of the disc is $h/2 \sim 10^{21}$ cm and $D \sim 10^{29}$ cm^2 · sec^{-1} (see [11]). In a halo with $R \sim 3.10^{22}$ for the same value of D, $T_h \sim R^2/2D \sim 10^8$ years. The source power in the disc model for the values indicated would be $U_d \sim W_d/T_d \sim 10^{42}$ erg/sec which exceeds the source power in the halo model by two orders of magnitude. It is important that to assure the presence of an injection of such power would be extremely difficult. Assigning a value of $T_d \sim 10^5$ years would be probably impossible also because of considerations concerning the chemical makeup of the cosmic rays and because of the degree of their anisotropy (for greater detail see [23]). Thus the disc model is inevitably tied in with the assumption that the cosmic rays be retained within the disc rather well, so that $T_d \gtrsim 10^7$. But even with an assumed layer of thickness $h \simeq 2.10^{21} \simeq 800$ pc which would correspond to the radio disc (see [33]) it would be quite difficult to guarantee a good trapping of the cosmic rays. Thus the value $T_d \sim 10^7$ years corresponds to an effective diffusion coefficient $D = ev/3 \sim 10^{27}$ which means that the effective mean free path would be $l \lesssim 0.1$ pc since the velocity of the particle motion along the field is $v \sim 10^{10}$ cm/sec. At the same time the distance between the clouds in the disc is $l_0 \sim 100$ pc and thus to guarantee a strong deflection of the particles along the path of $l \lesssim 0.1$ would be difficult.

However, the existence of a Galactic halo proves to be decisive here since when no halo is present where the halo serves as a region occupied by a field with $H \gtrsim 10^{-6}$, the cosmic rays cannot be retained in this region either.

III. 2. THE HALO PROBLEM

To a large extent, the name, "halo" is somewhat confusing because of the intermingling under the same term of two different concepts—the "physical halo" and the "radio halo" (see [34]). In the optical galactic disc, where $H \sim 10^{-5}$, a gas is distributed over a layer of thickness $h_g \sim 200$ pc with density $\rho_g \sim 10^{-24}$ g/cm^3. In intergalactic space in the boundaries of the Local group of galaxies, it is probable that $H \lesssim 10^{-7} \div 10^{-8}$ and $\rho \lesssim 10^{-28}$ g/cm^3. What is the nature of the transition between these two regions? The existence of a more or less sharp transition a priori is not very likely; the hypothesis concerning a "physical halo" thus simply reduces to the assumption that the transitional region has the dimensions $R \gg h_g \sim 3.10^{20}$ cm.

This type of assumption is natural both from the point of view of trapping the cosmic rays as well as from dynamical and other considerations (see [11, 35-37]). Moreover, one may consider the existence of a physical halo to be proven since the thickness of the observed radio disc in the Galaxy is $h = h_r \sim 2.10^{21}$ cm (see [33] and above). This fact, however, remains somewhat obscured due to the fact that the galactic halo is often identified with the radio halo whereas the latter term is understood as referring to a quasi-spherical radio emitting region of radius $R \sim 10$-15 Kpc $\simeq 3 \div 5.10^{22}$ cm surrounding the galactic disc. The question as to whether such a radio-halo indeed exists is still, strictly speaking, open. However, a large number of arguments in favor of the existence of a radio halo may be brought forth even at the present time. Thus, in the direction of the galactic pole the effective temperature of the radio emission of frequency $\nu \simeq 180$ megacycles is $T_f \simeq 100°$K. Using either this value or data known concerning other frequencies it is easy to see that the relativistic electrons whose concentration is observable in the vicinity of the Earth will yield radiation of this type if the electrons occupy a halo with $R \simeq 10$-15 Kpc having a field of $H \simeq 3.10^{-6}$ (see [21, 23, 29, 38]).

In so doing, the spectral emission index $\alpha = 0.5 \div 0.7$ gives rise to a differential electron spectrum having an index of $\gamma = 2d + 1 = 2 \div 2,5$. This does not contradict the data concerning the electron component in the vicinity of the Earth.

If, however, one were to assume that no radio-halo were present, than one would have to assume that the observed quasi spherical component of radio-emission is metagalactic. At the same time all of the evaluations that we know of give rise to the values $T_{f, Mg} \leq 20\text{-}30°$ for the metagalactic component for $\nu = 180$ megacycles/sec. Thus, rejecting the presence of a radio halo is equivalent to the completely unfounded assumption concerning the existence of a corresponding metagalactic radiation. However, the only reason one would have to cast doubt upon the existence of a radio halo would be the absence according to certain data of a clearly expressed angular dependence in the radio emission. (The asymmetrical position of the solar system would have to give rise for a halo that is symmetric with respect to the center of the Galaxy to a definite dependence in the radio emission intensity upon direction [33]). However, measurement of the angular dependence is complicated by a number of different circumstances, so that the true picture with regard to this question is unclear. Thus the existence of a radio halo has not, of course, been proven, but on the other hand it has by no means been disproven. Incidentally, existence of a radio halo with $R \simeq 3$ Kpc and $H \simeq 3.10^{-6}$ would not contradict even the above mentioned data concerning the dependence of intensity on direction.

By virtue of what we have mentioned, the existence of a radio halo in the vicinity of the Galaxy appears quite probable. (For the galaxy M31 the existence of a radio halo is accepted as proven, particularly for long waves).

In what follows it is particularly important to emphasize an additional factor. The assertion that no radio halo exists for the Galaxy would mean that there would be no quasi spherical region with $R \gtrsim 10$ Kpc acting as a source of radio emission with $T_f > 20\text{-}30°$ (for $\nu \simeq 180$ megacycles). But this would imply only a weak restriction on the parameters of the physical halo. In fact the intensity of the radio-emission is proportional to $H^{(\gamma +1)/2}$. If for $H = 3.10^{-6}$ the radio emission intensity of the halo were to correspond to a temperature of $T_f = 100°$, then for $\gamma = 2$ and $H = 10^{-6}$ we would then have $T_f \simeq 20°$ and radioastronomers would then conclude that there is no radio halo. At the same time the physical halo with $R \sim 10 \div 15$ Kpc and $H \simeq 10^{-6}$ still differs radically in its influence on cosmic rays from a metagalactic region with $H < 10^{-7}$.

In summary, it may be stated that there exists no data that could be used as evidence against the possibility of selecting a galactic model for the origin of cosmic rays in which use is made of the assumption concerning the existence of a physical halo with $R \sim 10\text{-}15$ Kpc. On the contrary the alternative quasi stationary model—the disc model, encounters difficulties and is considerably less probable.

III. 3. THE MOST PROBABLE MODEL

We thus consider the most probable model to be a galactic model with a halo. This model has been discussed several times since 1953(see [7]). We shall therefore introduce the parameter of this model without any detailed explanations.

The Galactic Model With a Halo

Radius $\quad\quad R \sim 3 \div 5 \cdot 10^{22}$ cm
Volume $\quad\quad V \sim 10^{68}$ cm^3
Energy of cosmic rays $\quad W_{cr} \sim 10^{56}$ ergs
Time necessary for particles to leave this system $\quad T \sim 3.10^8$ years

Power of the sources $\qquad U \sim \dfrac{W_{cr}}{T} \sim 10^{40}$ erg/sec

This scheme is quasi-stationary (the change in the mean intensity in the system is $\lesssim$ 10% during a period of $T \sim 1 \div 3.10^8$ years).

For the electron component: $W_e \sim 10^{54}$ ergs, $T_e \sim 10^8$ years, and $U_e \sim W_e/T_e \sim 3.10^{38}$ erg/sec.

The basic sources: Supernovae and, possibly, "small" explosions within the galactic nucleus.

In approaching the peripheries of the halo (for $R \gtrsim 10$ Kpc) the energy density w_G in galactic models must necessarily decay. For this reason the value $W_{cr} \sim w_G V$ may, for example, turn out to be equal to 3.10^{55} ergs. The lifetime T for protons and nuclei may also decrease but in the model under discussion we nevertheless have $T \gtrsim 10^8$ years. In this connection $U \sim 3.10^{39} \div 3.10^{40}$ are reasonable limits as such, reflecting the inexactness of the parameters. The characteristic lifetime of the electrons T_e is less than the time T, due to losses which increase with the energy. In the case where more exact computations are carried out, this fact would have to be taken into account.

The calculations related to this which we carried out were elucidated in § 17 in [11]. Since more recent calculations of this type in which up to date data is taken into account have not yet been completed, we shall confine ourselves here to making two comments. The calculations in [38] do not contradict our calculations if we take into account the fact that we made use of data concerning the intensity of nonthermal radio emission averaged over a half sphere in the direction of the galactic anticenter. This sort of intensity is about 2.5 times greater than in the galactic polar direction. As was already mentioned, in the case where the value $T_f = 100°$ is selected the electrons observed at the Earth yield the radiation that is required for a halo with the parameters $R \sim 10 \div 15$ and $H \sim 3.10^{-6}$. Calculations in [11] for secondary electrons would coincide with [39] for the contribution made by the proton component of the cosmic rays, provided that we take as the mean concentration of hydrogen within the complete volume of the Galaxy, including the halo, the value $n = 0.01$ cm^{-3}. The divergence between the results in [11] and [39] is due to the fact that in [39] a value for the density was used that was $1.5 \div 3$ times greater. We do not see a basis for such use. Moreover, the role of αp collisions,* was in [39], in our opinion, greatly exaggerated.

On the whole the data on the electron component appears to be in complete accordance with the model under consideration (see [5, 11, 21, 23, 29, 34, 38-40]. The same may also be said concerning all of the data that we are familiar with on the proton-nuclear component. In a number of cases, however, the experimental values are imprecise and the calculations are either inexact or insufficiently determined. For this reason one can still not regard the halo model to be proven especially when one considers the state of the halo problem itself. In future research it would be extremely important to elucidate the dimensions of the halo. One might expect that the solution to this problem would be attainable by radio—methods (separating the radiation component corresponding to the radio halo). If it turns out in this manner that the fundamental assumption of the model are valid, then our attention would be focussed basically upon quantitative calculations (chemical composition, secondary electrons, and positrons, γ rays, anisotropy)

*It is assumed in [39] that for collisions between a proton and a helium nucleus the interaction takes place with a single nucleon of the nucleus. However, in contradiction to this position it is generally accepted that the energy of the mesons generated (for the same energy per nucleon for the incoming particle) is four times greater for αp collisions than for pα collisions.

and their comparison with observed data, as well as upon the investigation of the sources of cosmic rays themselves.

III. 4. THE SOURCE PROBLEM

The model chosen, namely, the galactic halo model, imposes definite conditions upon the sources of cosmic rays. First and foremost it fixes the power of injection. However, the sources themselves cannot yet be considered to be uniquely determined. In other words, for a sufficiently complete description of the model it would be necessary to make the choice of sources more precise as well as to develop in the future a theory of sources. The problem of sources has been discussed many times. In these discussions the main sources were at first considered (see [1-4, 7-9]) to be bursts from supernovae. In recent years the possibility was pointed out that "small" explosions in the galactic nucleus (see [5, 6, 11, 20, 23]) might also play a substantial role. This opinion appears to prevail even at the present time.[†]

According to data [41] supernovas flare up in the Galaxy once every fifty years = $1.5 \cdot 10^5$ seconds on the average (see also [46]). For this reason the mean energy that would have to be converted into cosmic rays during the flare of a single supernova would have to be $W_{sn} = 1.5.10^9 \, U \sim 10^{49}$ ergs (where the power of the entire injection is $U = 10^{40}$ erg/sec). As we know a value of $W_{sn} \sim 10^{49}$ is quite possible.

The existence of "small" explosions within the galactic nucleus may be considered to be quite likely, not only by virtue of comparison with other galaxies but also by virtue of data dealing with motion in the central region of the Galaxy [42]. A mean power of injection of $U \sim 10^{40}$ erg/sec may be assured, for example, if during a single explosion an energy of $W_n \sim 3.10^{54}$ ergs $\sim M_\odot c^2$ is converted into cosmic rays and if such explosions repeat themselves every 10^7 years. A galactic nucleus has not long ago been discovered in infrared region (see [43]) and the total power (luminosity) of the source is 8.10^{40} erg/sec. This luminosity is approximately 2.10^7 times greater than the luminosity of the sun. Due to a number of considerations one may expect that the mass of the nucleus would be $M \gtrsim 3.10^7 \, M_\odot$. The explosion of the galactic nucleus would correspond to a loss in stability or even a collapse. A conversion into cosmic rays of an energy of $W_n \sim 3.10^{54}$ ergs even under conditions of very low efficiency in acceleration would in this case be quite possible (it suffices to mention that for a nucleus, whose mass is as we indicated above $Mc^2 \gtrsim 5.10^{61}$ ergs; for some quasar models the energy release reaches $10^{-3} \, Mc^2$). If the main part of the observed cosmic rays were accelerated as a result of explosions in the nucleus (in order that there accumulate an energy of $W_{cr} \sim 10^{56}$ there would clearly have to take place 30 explosions having $W_n \sim 3.10^{54}$ ergs), then one may expect a certain degree of instability, i.e. variations in intensity. However, if the variation is less than 3% as in the case discussed above, then it is unlikely that it would be detected. The same thing may also be said about other methods that are known.[**]

[†]Novae stars as well as X-ray sources of type Sco XR-1 do not yield any noticeable radio emission. Thus there is no particular bases for considering these objects to be powerful sources of cosmic rays although in principle it is possible (see [11, 23]). About the role of polsars see Ref. 48.

[**]What was said here does not refer to electrons of high energy. If the last explosion in the nucleus of the Galaxy took place 10^7 years ago (a smaller value would be extremely improbable), then the electrons thus accelerated would not at our time be able to have any energy greater than $E_{max} \simeq 3.10^{10}$ ev/cm (see (9) for the case where $w_T + H^2/8\pi \simeq 2.10^{-12}$ erg/cm3 and $\tau = 3.10^{14}$ sec). If we set

It is thus clear that to separate the contributions from the supernovas and from the explosions in the nucleus would not be easy. It should be mentioned, however, that a high degree of effectiveness among supernovae as injectors of relativistic parti- cles has already been established; with respect to explosions in the nucleus of the Galaxy, at the same time, there prevails a complete lack of definiteness in this re- gard. For this reason supernovae are still the most likely candidates for the role of main source of cosmic rays in the Galaxy.

Conclusion

During the period since the last London Cosmic Ray Physics conference in 1965 no data of a completely definite character has appeared that would cause us to change our estimate of the general state of the problem on the origin of cosmic rays. Significant progress has been achieved, however, with regard to the electron component of radiation. If residual metagalactic thermal radiation with $T \simeq 3°K$ exists, and it would be quite difficult to doubt this in spite of the lack of measurements for the wave lengths $\lambda < 1.5$ cm, then the uniform metagalactic model for the origin of the electron component would have to be ruled out. Local metagalactic models would still not have to be completely ruled out; however, severe restrictions would necessarily be imposed on them in this case, particu- larly in the case of electrons with energy $E > 10^{10}$ ev. The assumption that the sources of the electron and proton-nuclear components of cosmic rays are dif- ferent (for energies that are not too high) would appear to be rather improbable even from general considerations. This, combined with what was mentioned ear- lier, is seen by us as one of the basic arguments against the metagalactic model for the proton-nuclear component. Another argument no less important is based on energy considerations. Within the framework of evolutionary cosmology the uni- form metagalactic model appears to us as unacceptable even from these (energy) considerations alone (see however [45]). The same may be said, although in a less definite manner, about the "Centaurus A model" and about certain other local mod- els. Strictly speaking a local metagalactic model or a nonstationary galactic model appears possible only under the assumption that in the nucleus of the Galaxy or in the nuclei of the closest galaxies giant explosions occurred not long ago ($T \lesssim 10^8$ years). (For Centaurus A the situation involves an injection of cosmic rays with an energy of $W_{cr} \sim 10^{63}$ ergs, for which a mass of $M \gtrsim 10^{12} M_\Theta$ must collapse). We do not see any real basis for such hypotheses. Even if it were to turn out that quasars have a local nature, this would still not give any direct guarantee with respect to the validity of the local metagalactic model or of the nonstationary galactic model for the origins of cosmic rays. However, at the moment we see no real arguments in favor of a local model for quasars.

Among the galactic models the most probable one is the halo model. We do not see any contradictions or difficulties in this model, but it still has not been proven and in many respects it has not been quantitatively worked out enough with regard to most recent data. *

$H \sim 10^{-5}$ for this case which is a more probable estimate, then $E_{max} \simeq 10^{10}$ ev. In the above, the travel time for the electrons moving from the nucleus to the Earth was not taken into account. For diffusion propagation with $D \sim 10^{29}$ this time is $T \sim R^2/2D \sim 10^8$ years, and $E_{max} \simeq 1 \div 3. 10^9$ ev (clearly this argument is also valid for the case where the nucleus generates electrons all the time).

*An attempt at constructing a quantitative theory for the origin of cosmic rays was undertaken in our book [11]. We have now begun to prepare a second edition of this book and we intend to review all of the computations appearing therein in the light of new information.

By virtue of what has been said above, it is clear that it will be necessary to return to the same questions concerning metagalactic cosmic rays, the galactic halo, and sources of cosmic rays even in the future. The introduction of conclusive clarity as well as the proof of the validity of a number of opinions within this series of problems can be attained only with great difficulty and will unfortunately demand a long period of time before they can be resolved. We submit, however, that the history of research on the origin of cosmic rays during the last fifteen years does not provide a basis for skepticism in this regard. On the contrary it bears evidence to real progress.

BIBLIOGRAPHY

1. V. L. Ginzburg, Memoria del V. Congreso International de Radiacion Cosmia p. 546 Mexico (1958); Nuovo Cimente, Suppl. 3, N 1, 38 (1956).
2. V. L. Ginzburg. Nuovo Cimento. Suppl. 8, N 2, 430 (1958).
3. V. L. Ginzburg, Preceedings of the international conference on cosmic rays, Moscow 3, 200 (1960) (In Russian). There are also English versions.
4. V. L. Ginzburg and S. I. Syrovatskii. Progress Theoret. Phys. Suppl. N 20, 1 (1961).
5. V. L. Ginzburg and S. I. Syrovatskii. Proc. Intern. Conf. Cosmic Rays, India 3, 301 (1963).
6. V. L. Ginzburg and S. I. Syrovatskii. Proc. Intern. Conf. Cosmic Rays, London 1, 53 (1965); Uspekhi Fiz. Nau. 88, 485 (1966).
7. V. L. Ginzburg, Soviet Physics--Uspekhi Fiz. Nauk. 51, 343 1953 (In Russian). See also Fortschritte der Physik 1, 659 (1954).
8. S. Hayakawa, K. Ito, Y. Terashima. Progress Theoret. Phys. Suppl. 6, 1 (1958); see also. Suppl. N 30, (1964).
9. V. L. Ginzburg. Progress in Elementary Particle and Cosmic Ray Physics 4, chap. 5, Amsterdam (1958).
10. P. Morrison. Handbuch d. Physik. 46/1, 1 (1961).
11. V. L. Ginzburg and S. I. Syrovatskii. Origin of cosmic rays. Pergamon Press (1964).
12. V. L. Ginzburg. Astronautica Acta. 12, 136 (1966).
13. G. R. Burbidge and E. M. Burbidge. Limits to the distances of the quasi-stellar objects deduced from their absorbtion line spectra (to appear in Ap. J. Letters, May. 1967).
14. J. Terrell. Science, 154, 1281 (1966).
15. V. L. Ginzburg, Astron. Zhar 42, 1129 1965 (In Russian) This journal is translated into English. Sov. Astron. AJ
16. I. Lerche. Ap. J. 147, 689 (1967).
17. G. Puppi, G. Setti and L. Woltjer. Nouvo Cim. 45, 252 (1966).
18. J. E. Felton and Ph. Morrison. Ap. J. 146, 686 (1966).
19. Texas conference on relativistic astrophysics. New York, (January, 1967).
20. G. Burbidge, Sci. American 215, N 2, 32 (1966).
21. W. R. Webber a. Ch. Chotkawski Determination of the intensity and energy spectrum of extra-terrestrial electrons in energy-range 70-2000 MeV. Preprint (1966).
22. S. I. Syrovatskii, JETP (In Russian) 50, 1133 1966, Soviet Astronomy—AJ 43, 340 1966
23. V. L. Ginzburg and S. I. Syrovatskii. Proc. IAU Symposium N 31, paper N 71, Ed. by H. Von Woerden. Academic Press (1967)
24. G. R. Burbidge. Progress Theoret. Phys. 27, 999 (1962). G. R. Burbidge and F. Hoyle. Proc. Phys. Soc. 84, 141 (1964).

25. M. Schmidt, Proc. IAU Symposium N 31, Ed. by H. von Woerden Academic
     Press (1967).
26. M. Schmidt, Ap. Journal 146, 7 (1966).
27. C. M. Wade. Phys. Rev. Letters 17, 1061 (1966).
28. G. H. Macdonald, A. C. Neville and M. Ryle. Nature 211, 1241 (1966).
29. H. Okuda and T. Tanaka. The galactic magnetic field derived from cosmic-
     ray electrons. Preprint (1967).
30. R. R. Daniel and S. A. Stephans. Phys. Rev. Letters 17, 935 (1966).
31. W. L. Kraushaar and G. W. Clark Phys. Rev. Letters 8, 106 (1962). Proc. Inter.
     Conf. on cosmic rays, India (1963).
32. L. I. Dorman, O. I. Inozemtseva, E. A. Mazaryuk and Z. I. Soloveva, Geomag-
     netism and Aeronautics, 7, 23, 1967 (In Russian) Also translated.
33. J. E. Baldwin. Proc. IAU Symposium N 31, paper 56, Holland (1966).
34. V. L. Ginzburg. Proc. IAU Symposium N 31, paper 61, Holland (1966).
35. S. B. Pikel'ner, Soviet Physics--Doklady 88, 229 (1953) (In Russian).
36. E. N. Parker, Ap. J. 142, 584 (1965). Proc. Inter. Conf. on Cosmic Rays.,
     London, 1, 126 (1965).
37. J. H. Oort. Structure and evolution of the galactic system. Transactions IAU
     12a, 789 (1965).
38. J. E. Felton. Ap. J. 145, 589 (1966).
39. R. Ramaty and R. E. Lingenfelter. J. Geophys. Res. 71, 3687 (1966).
40. S. D. Verma. Phys. Rev. Letters 18, 253 (1967).
41. Sky and Telescope. 33, 3 (1967).
42. Proc. IAU Symposium N 31, Ed. by H. von Woerden Academic Press (1967).
43. Sky and Telescope. 33, 203 (1967).
44. R. Bergamini, P. Londrillo and G. Setti, Nuovo Cimento 52B, 495 (1967).
45. V. L. Ginzburg. Astrophysics and Space Science. 1, 125 (1968).
46. P. Katgert and J. H. Oort. Bull. Astron. Inst. Nederlands 19, N 4, 239, (1967).
47. V. L. Ginzburg and S. I. Syrovatskij. Astrophysics and Space Science 1,
     442 (1968).
48. V. L. Ginzburg, V. V. Zheleznyakov and V. V. Zaitsev. Astrophys. and
     Space Sci. (in press).

For ref. 23, 25, 33, 34, 42, more complete reference is: Radio Astronomy and
the "Galactic System". IAU Symposium N 31, Edited by H. von Woerden
Academic Press, London and N. Y. (1967).

INDEX

INDEX

61